Pseudoscience

Pseudoscience

A Critical Encyclopedia

Brian Regal

GREENWOOD PRESS
An Imprint of ABC-CLIO, LLC

Santa Barbara, California • Denver, Colorado • Oxford, England

Library of Congress Cataloging-in-Publication Data

Regal, Brian.
   Pseudoscience : a critical encyclopedia / Brian Regal.
      p. cm.
   Includes bibliographical references and index.
   ISBN 978-0-313-35507-3 (hardcopy : alk. paper)—ISBN 978-0-313-35508-0
(ebook) 1. Pseudoscience—Encyclopedias. I. Title.
   Q172.5.P77R44   2009
   001.9—dc22   2009023703

13 12 11 10 09   1 2 3 4 5

This book is also available on the World Wide Web as an eBook.
Visit www.abc-clio.com for details.

ABC-CLIO, LLC
130 Cremona Drive, P.O. Box 1911
Santa Barbara, California 93116-1911

This book is printed on acid-free paper ∞
Manufactured in the United States of America

Contents

List of Entries

Acknowledgments

I would like to thank Kevin Downing at ABC-CLIO for bringing this project to me. Also thanks to Alexandra Bolusi, Jessica Rajs, Scott Heeney, Maria Gatto, Jillian Martinez, and Timothy Hulme for their research help.

Introduction

A pocket full of mumbles, such are promises . . .
—Simon & Garfunkel, "The Boxer" 1968

A book on pseudoscience, especially an encyclopedia, can be a tricky proposition. The problem rests in the realm of definitions. Just what is pseudoscience? Is it cultural, social, theological, political, or all of the above? What science is and what it is not is a crucial question in the 21st century. Making uninformed choices or choices made based upon faulty beliefs or dubious logic can have dire consequences. Choosing pseudoscience over science can stifle the economy, confuse lawmakers, cause strife in our schools, breed hatred, or simply kill us all so we better know what we are talking about or at least have a way of discussing it.

The study of pseudoscience is a study in exclusivity. Many of the topics covered in this book are researched, believed in, and promoted by small communities. They often see themselves in opposition to mainstream science in a consciously held "us versus them" mentality. Taking a confrontational position against an ill-defined opposing force helps give the community a sense of cohesion and shared purpose, especially if the practitioners are amateurs or if they are professional scientists whose work is not well regarded by their colleagues. They see any and every rejection by the mainstream as proof that the other side is out to get them and trying to squash their revolutionary ideas. It allows them to claim an intellectual high ground by arguing that it is the amateurs who are more scientific than the professionals because they are open to new ideas and are not stultified by hierarchies and preconceived notions, the way they presume professionals are. They are not completely off the mark, however. Researching anything too far, or in some cases, even a little bit, beyond the norm can result in being labeled fringe science or pseudoscience. Such a label is often used derogatorily to insult or make something seem foolish. Sometimes these pursuits deserve such a label, and sometimes not.

A working definition of pseudoscience (from the Greek for false science) is that it is any intellectual or technological pursuit that purports to use scientific methodology or philosophy to study or prove some temporal or physical reality, but which in fact does not (for example, using a temperature gauge to check for the presence of ghosts). Using a temperature gauge is an accepted technique to determine how warm or cold a particular environment is. This is genuine science because years of experience, testing, double checking, and experimentation have shown that mercury rises and lowers uniformly in a tube depending upon the temperature around it.

Believers argue that ghostly spirits cause temperature changes when they appear. On the face of it, this sounds like a logical way to look for a ghost, an almost foolproof approach. If you think a ghost is nearby, test the air with a temperature gauge. If fluctuations in temperature register, you have a good idea that a spirit is present. Practitioners of pseudoscience rarely reject science, they just have their own unique ways of employing it. Unfortunately, there is no accepted evidence that spirits or ghosts exist—nor even a uniform definition of what a ghost is—so checking the temperature is an empty exercise. However, there is as of yet no way of knowing if ghosts change the temperature of their surroundings, even if they do exist, so while this procedure seems very scientific and straightforward, using a temperature gauge to find ghosts falls into the realm of pseudoscience. It should be noted, however, that someday in the future it may be determined that ghosts do indeed exist and that they do effect the air temperature around them; this would make the temperature gauge technique a legitimate, and scientific, way of monitoring them. At that time, ghost hunting would no longer be considered a pseudoscientific exercise.

The world of pseudoscience does not often check itself. When a scientist or historian has an idea, she brings the idea to her community in the form of a peer review. This can be a brutally humbling process. Others who specialize in the field are given a chance to read and critique the scholar's ideas. This is a process designed to weed out poorly constructed or argued ideas, to help scholars improve, and to help the author to get a better sense of what she is doing and force a confrontation. Ideas that go through this and survive—and certainly more fail than survive—will eventually be accepted by the wider scientific community. This is what Thomas Kuhn called the "paradigm model of science." The community fights over concepts until they are satisfied that they are logical and worth consideration. Pseudoscience enthusiasts take this standard operating procedure and see it as stifling, retardant to new ideas, even sinister, especially if their ideas are the ones that get rejected.

So, what is science? A working definition of science is that it is an intellectual exercise in which knowledge about the workings of the universe is looked for. This is done according to rules of inquiry. Empirical evidence must be gathered, quantitative analyses completed, experiments performed to check a hypothesis, and questions constantly asked and ways are looked for to show the idea is false. The idea of a scientific method was first put forward by the British philosopher Francis Bacon (1561–1626). He argued that techniques ought to be used to determine the laws of nature, rather than the haphazard approach that had previously dominated scholarly work. Unsatisfied with medieval scholasticism's reliance on past authorities, Bacon put forward suggestions for methods for using only empirically and observationally based data—thus rejecting past authorities—for determining the laws of nature.

While the story is far more complicated, and pure Baconian empiricism is rarely used on its own in modern scientific practice, Bacon is often credited with helping to inaugurate the modern scientific approach. A much later addition to how science is approached came when Karl Popper (1902–1994) added the notion of falsifiability to the simplistic observational approach. Calling his overall philosophy critical "rationalism," Popper argued that scientific theories could only be argued in the abstract and that simple experimentation cannot always prove or disprove a theory. A theory, unlike the popular understanding of the concept, is not a simple guess, rather it is a supposition based upon a great deal of evidence that helps explain things and answer questions about some broader field. The possibility to prove an idea wrong is what separates science from non-science. If one argues that gravity is a force that tends to draw objects toward the center of a mass, one can test that by letting go of an object. If it falls, you can say there is evidence for gravity, if it does not, then perhaps gravity is not what it is thought to be, or maybe the body is not massive enough to generate gravity. A genuinely scientific idea, Popper claimed, could possibly be proven wrong while a pseudoscientific one could not: How does one disprove that space aliens have come to earth and kidnapped humans for experimentation?

In the 1960s, Thomas Kuhn (1922–1996) put forward the paradigm concept. He argued that an intellectual community held basic ideas and theories about how things worked, called paradigms. These ideas were the foundation of all knowledge of that field of inquiry, and must be accepted by the community in order for progress to be made. An example of a paradigm is a sun-centered solar system, or heliocentric cosmology. That the earth orbits the sun is a basic piece of information that must be accepted in order to understand how the solar system works. If you don't accept it, little about planetary mechanics will make any sense. The same holds for evolution. Nothing about biological function on earth makes sense unless you accept evolution, therefore evolution and a heliocentric view of the universe are considered paradigms. The tricky problem with science and pseudoscience is how to determine just where science ends and pseudoscience begins. At the ends of the spectrum it is easy, it is in the middle ground where this is more problematic.

The demarcation between science and pseudoscience is a topic of great import to scientists, historians, and philosophers of science, and there is a considerable body of scholarly literature on the topic that readers can easily access. Thomas Gieryn argues that pseudoscience is an idea used by science in a symbiotic way to show where its own edges and boundaries are. Where most academics who write on pseudoscience are concerned with how to determine where the separation is, philosopher of science Paul Feyerabend (1924–1994) famously argued that there is no way to tell the difference between science and pseudoscience. He claimed there were no stable sets of rules for how science is done, and thus no boundaries could separate it from any other intellectual endeavor. His anarchistic attitude drew many supporters and as many critics. In *Against Method* (1975), he said that there was no one "scientific method" that all practitioners of science follow, and claimed that most great intellectual advances were made not due to formal steps but on account of ideas and procedures outside the normal techniques of scientific inquiry. Being forced to follow rigid rules was, he said, what kept creative thinking from happening, and creative thinking is what pushes science forward. Some practitioners of pseudoscience have used this same logic, though not as elegantly or logically, to support not following the rules of scientific and intellectual inquiry.

Another philosopher of science, Paul Thagard, who thought one does need rules and structures, used astrology as an example of what pseudoscience is in a 1978 paper. He argued that there were criteria one could employ to determine what was or was not pseudoscience. Astrology was pseudoscience, Thagard said, because those who practice it do not attempt to develop or advance the theory (modern astrologers have adapted computer modeling to the construction of charts, but the underlying theory has remained static since at least the Middle Ages), they are not critical of their own work, show no concern for attempts to evaluate the theory in relation to other disciplines such as astronomy, and are often selective in what evidence they put forward to the wider community. He noted that astrologers also tend to gear their work toward resemblances rather than established facts. In other words, if something looks right, it must be. This is the old "if it walks like a duck and quacks like a duck, it must be a duck" argument. Science rejects this because some things do look like ducks yet are not ducks. Robert Weyant argues that just because something is science done incorrectly, it does not mean it is pseudoscience, but it doesn't mean that it is science, either. There may be various grades of non-science of which pseudoscience is only one. If an idea or belief has no way of being tested, he argues, then it cannot be judged one way or another.

Science generally falls under the broader category of naturalism, sometimes called materialism. While there are a number of definitions of naturalism/materialism in different fields of inquiry, there are two types of scientific naturalism. Methodological naturalism attempts to explain how the universe works without any reference to metaphysical or theological explanations. While methodological materialists might acknowledge that supernatural elements could exist in the universe, they avoid using them in any explanation of phenomena, arguing that science has no tools by which to prove or disprove the metaphysical, therefore all scientific explanations have no choice but to rely upon rational and natural evidence. Philosophical materialists not only avoid reference to the supernatural, they reject the notion that it even exists. They argue there is nothing besides the material world, and that only material or natural explanations can be used. The materialist worldview, as one can imagine, is roundly rejected by theologians and some fans of the fantastic. The materialist approach accounts for why so many individuals in the world of pseudoscience work so hard to bring scientific methodologies to bear upon their problem. This is their saving grace, because if one wants to prove that a perpetual motion machine is possible, one will have to build it; if a person wants to prove he can communicate with the dead, this, too, must be reproducible in experiments over and over again and in the same way each time. The possibility that these ideas can be proven to be genuine drives much of the research covered in this book. Lone believers and associated groups of believers, in their attempt to prove the existence of Bigfoot or to show that space aliens visit our planet or that people have had past life experiences, are out in the field right now trying new machines and new techniques. This is how pseudoscience can sometimes evolve into real science.

This brings us back to Thomas Kuhn and paradigms. Kuhn argued that accepted ideas were not written in stone; they can be overthrown in a paradigm shift. This only happens, however, if an idea has an enormous amount of evidence backing it up, if it answers questions the previous paradigm could not, and if the new idea has the support of the majority of the community. As a result, paradigm shifts happen infrequently, but they do happen. So, if a scientist wants to prove ghosts exist, she will need more than creepy sounds recorded at night and orbs caught in photographs. When enough evidence to support the existence is accepted, and when the existence of ghosts answer other questions about the universe, then a paradigm shift may take place.

While pseudoscience is usually associated with untrained experimenters and researchers, scientists themselves can sometimes fall into the pseudoscience trap. It has been argued that some forms of commercial product testing on animals is pseudoscientific, as it does not really generate information useful to determining whether or not those products adversely affect humans. Also, if a scientist accepts or refuses to accept the possibility of something without properly examining the evidence, that too is pseudoscientific. Refusing to acknowledge an idea like global warming because it conflicts with a political position or adversely affects a corporate employer would also be pseudoscience, as it is based on something other than empirical evidence. Science is supposed to be about the cool and rational analysis of evidence. Whenever that is absent, the exercise can be considered pseudoscience. This leads us to another conundrum.

This book is meant to get the reader thinking about what pseudoscience is and to question whether the various topics included here are worthy of the name. While it takes the form of an encyclopedia, this book is meant to be more than just a list of explanations: It is an extended discussion of what pseudoscience is. While most of the categories here are pseudoscientific, not all of them are, or at least they should be rethought. There is also a question about the term itself. Does the term "pseudoscience" help explain anything? Does it force us to reconsider the boundaries between what is science and what is not? Are the lines drawn just where they should be, or do they need to be realigned? Is the definition of pseudoscience itself pseudoscientific? These are the questions this book is designed to prompt the reader to consider. Ironically, the book also should not be taken as encyclopedic. The topics herein are not addressed as a definitive compendium of dry names and dates, rather they are discussed with an eye toward figuring out where they fall within the spectrum of what science is supposed to be.

These questions are important because some of the topics covered in this book exist in a twilight world where they could be classified as fringe science, the occult, religion, conspiracy theory, or as something else. This is evidence of the porous condition that can sometimes exist at the boundaries of ideas. Something thought of as on the fringe today may be deep inside the mainstream tomorrow. The history of science is full of examples of ideas or phenomena once thought to be pseudoscience but which are not thought of as such now. Plate tectonics, the heliocentric model of the universe, quantum mechanics, herbal medicine, even evolution and electricity were once thought dubious concepts with little or no rational evidence to support them. Conversely, ideas like racial theory and phrenology, medical practices like lobotomies, bleeding, and electroshock therapy, all once thought to be good science, are now considered pseudoscience.

In the mid-20th century, smoking cigarettes was promoted as a healthy practice, as was eating lots of red meat, two things we now know to be supremely unhealthy.

In addition to listing, explaining, and critiquing these topics, there are a number of underlying issues here. Along with the conflict of science and pseudoscience, there is the relationship between reason and belief, between science and religion, professional scientists versus amateurs, the role of scientific authority and how science and history are perceived by the public, and why it is so important to keep science, pseudoscience, and religion apart. Do we actively reject any and all knowledge if it has a whiff of pseudoscience about it? Do we, like Paul Feyerabend argued, accept all ideas, no matter how anarchistic they may seem, with an eye toward believing that all knowledge, even false knowledge, has something to tell us? These are questions the reader must answer for him- or herself.

To study pseudoscience, one must ask lots of questions. We cannot take phenomena at face value. We must try to avoid preconceived notions about what we are looking for (this may be the hardest part), and we must be able to prove our contentions in the bright light of day. We must discard metaphysical notions and avoid falling back on theological beliefs to explain things we cannot otherwise explain, and have to be ready to be told that what we have done is not good enough yet and that we need to go back and work harder. We must learn to think like scientists, and think of how this knowledge and its uses affect society.

In 1940, as the Nazi juggernaut was conquering Europe and the Japanese Empire's military machine was rampaging across Asia, a group of scientists and theologians met in New York for a conference on science, religion, and the state of the world. They declared that they were in opposition to the irrational pseudoscientific belief systems of the Nazis and Japanese. Race theory, conspiracy theory, and the perversion of and inappropriate mingling of science and religion for twisted political ends was something that had to be opposed, they argued, or it would lead the world into totalitarian subjugation or plunge it into oblivion. The reliance on pseudoscience and religious fanaticism in Germany and Japan had "introduced intellectual confusion in their educational systems, in their literatures, and in organs of public opinion generally," the group argued. Reasonably thinking people of all groups had to combine their efforts to keep it from happening at home. In the early 21st century, the threat has yet to subside; we still need to keep our guard up. We still need to keep thinking. Having made this dire connection, it is fair to mention that, while we should always be wary of outlandish claims or discoveries meant to support political or religious ends, or which just seem too fanciful to work, we should also be prepared not to always dismiss such claims out of hand without an examination. Traveling down the dark road of the intellect along tracks others avoid, in the rain against all advice from others, often leads to running into a wall, a ditch, or over a cliff's edge to destruction. Occasionally, however, driving down the dark, lonely road can lead to the bright light of new lands.

There is one subject heading in this book that I have fabricated entirely myself. I will personally send a free, signed copy of this book to the first person to identify the entry correctly and contact me directly with the answer.

Further Reading

Bacon, Francis. 2008. *Francis Bacon: The major works*. Oxford: Oxford University Press.

Davis, Watson. 1940. Science, philosophy, religion find grounds for common front. *Science News Letter* (September 21): 180–90.

Gieryn, Thomas F. 1983. Boundary-work and the demarcation of science from non-science: strains and interests in professional ideologies of scientists. *American Sociological Review* 48:781–95.

Kuhn, Thomas. 1962. *The structure of scientific revolutions*. Chicago: University of Chicago Press.

Popper, Karl. 2001. *All life is problem solving*. New York: Routledge.

Thagard, Paul R. 1978. Why Astrology is Pseudoscience, Proceedings of the Philosophy of Science Association, Volume One: Contributed Papers, 223–34.

A

2012, THE YEAR

The belief that, according to an ancient Mayan calendar, the world will end in the year AD 2012. This interpretation of a Mayan text has excited worldwide interest and legions of Web sites and blog discussions. New World archaeologists, anthropologists, and scholars in the field of Mayan cultural studies and history say that the growing excitement is misplaced because it is based upon faulty but popular misreading and misinterpretations. Mayan civilization broke time into blocks of 394.5 years. Each time block is called a *bak'tun.* The latest *bak'tun* will end in 2012. Mayan metaphorical language about the ending of time has been interpreted in popular readings as meaning the world will end. This reaction is similar to the near hysteria generated over the year 2000, when doomsayers claimed the Christian millennium or end of times would come to pass on that date. There are many examples of predictions of the end of the world worldwide and across time. As of this writing, all have proven spurious and ill-founded.

ACUPUNCTURE

Acupuncture (needle piercing) is a traditional Chinese medical treatment that incorporates the use of fine needles driven into the skin to stimulate any of the 800 specifically designated points on the body. This painless practice is meant to regulate energy flow, or *"chi,"* within the body. Body function and immunity are said to be increased with this method. Other elements of Chinese medicine, such as herbalism, massage, moxabustion, and diet are incorporated into acupuncture treatments. Acupuncture can be used to treat an endless variety of illnesses and is said to have been used consistently for over 3,000 years.

The *chi,* or body energy, is said to affect every level of the body's functions. The belief is that, with the proper flow of *chi,* the body will be adept at healing and balancing itself. Disease and ill health are therefore said to be the products of blocked or stagnant *chi* that was unable to travel freely through the meridians of the body. A well-trained acupuncturist will recognize where the weakness or blockage within the meridian lies. The symptoms are used as a clue to where the pattern of disharmony originates, and needles are then used to manipulate and correct the flow of energy. The invisible meridians run through every organ in the body, and each point along a specific meridian can be affected if there is disharmony at other points. For example, because the teeth and the stomach are on the same meridian, teething babies often suffer from digestive disorders. Another practice also based on Chinese traditional medicine is acupressure. Like acupuncture, acupressure is said to stimulate and smooth the flow of the body's *chi* by using the pressure of the fingertips instead of needles. Massage is incorporated in this method to encourage the *chi*'s flow through the body's energy channels.

Critics charge that acupuncture as it is practiced today does not have an ancient pedigree. Earliest Western reports of the technique date from the 1680s, but do not describe the meridians or *chi.* Further, the needles described in the earliest reports were rather large and inserted into a woman's womb. Acupuncture's first appearance in North America was in 1826, when it was used to revive drowning victims. The Chinese government banned acupuncture in the mid-20th century, but it was brought back under the Communists as an example of acceptable, non-Western, non-decadent people's medicine (Mao's government was also responsible for coining the term "Traditional Chinese Medicine"). Some Western doctors today argue that acupuncture is pseudoscience because there is no way to conduct blind studies of it and reports of its efficacy are unsubstantiated and inconsistent.

See also: Alternative Medicine; Blind Testing.

Further Reading

Hall, Harriet. 2008. "What about acupuncture?" *Skeptic* 14 (3): 8–9.
Xinnong, Cheng. 2005. *Chinese acupuncture and moxibustion.* Beijing: Foreign Languages Press.

AGE OF ENLIGHTENMENT

Also known simply as the Enlightenment, this was a period of intellectual and philosophical growth beginning in 18th-century Europe and spreading rapidly around the Western world. As part of a complex movement that defied simple codification, Enlightenment philosophers sought to promote the primacy of reason and rational thought over the prevailing belief in the supernatural as an explanation for everything. Some tended to reject past authorities for an emphasis on empirically gathered data while others did not. Some Enlightenment *philosophes* were antireligion, and several were outspoken opponents of organized religion and its inherent hypocrisy, and against religion as a concept itself, but not all. This period is considered the birth of modern materialist science. In addition to the reconfiguring of the natural sciences, the Enlightenment also stressed the primacy of individual rights over those of the state and the tenets of democracy over monarchy. Political philosophers of this period studied the nature of the relationship between the individual and the state. In this way, the Enlightenment also saw the beginning of the concept of political science. The framework of modern science and democracy are based upon Enlightenment ideals. Those who reorganized the British and French governments during the 18th century were inspired by Enlightenment principles. The Founding Fathers of the United States also borrowed heavily from the Enlightenment to forge a new nation. Indeed, the U.S. Constitution is an Enlightenment document, not, as some claim, a biblically based one. The impact of the Enlightenment on modern science and politics cannot be overestimated. The concepts and suppositions of the Enlightenment are roundly disparaged by totalitarians, religious fundamentalists, political conservatives, antiscience, anti-evolution, and intelligent design proponents.

Further Reading

Dupre, Louis. 2005. *The enlightenment and the intellectual foundations of modern culture.* New Haven, CT: Yale University Press.

Hill, Jonathan. 2004. *Faith in the age of reason: The enlightenment from Galileo to Kant.* Downers Grove, IL: InterVarsity Press.

ALCHEMY

A protoscience that involved research into the refining of salts, metals, paints, and medicines as well as the transformation of common, lesser matter into rarer and more valuable forms. Though it reached its height of popularity in the late Medieval and Renaissance periods, the discipline has been practiced in one form or another from the ancient world to the present day. The word alchemy has its roots in the Greek *chemeia,* itself a derivation of the word for smelting metals. When such practices entered the Islamic world in the early Middle Ages, it acquired the Arabic definite article *al,* producing *al-kimiya.* It is from this linguistic construction that the modern English words alchemy and chemistry derive.

The practice of alchemy combined the studies of metallurgy and chemical processes through laboratory experimentation with the studies of medicine and disease and mysticism, religion, and philosophy. The most well-known aim of alchemy was to find a technique through which the alchemist could transmute base metals like lead into gold. For some alchemists "the great work," as they called it, was less about the prosaic transformation of materials than about a higher transformation of one's soul: to leave baser human foibles behind for spiritual enlightenment. The vehicle by which either transmutation would occur was the production of the Philosopher's Stone. Small portions of this material added to lead were said to change it into gold. Taken internally, the Philosopher's Stone would act as an elixir of life. The transformative aspect of alchemy was only one part of a larger intellectual enterprise.

The primary hurdle in the study of the history of alchemy is answering the questions of just what was alchemy and what are its murky origins? It seems to have been practiced in numerous forms in different cultures and times around the world, from the Middle East to China and India. One result is that there is no one explanation—or panacea, to use an alchemical term—to cover the wide range of diversity of beliefs and objectives that fall under the category of alchemy. The common popular view of alchemy, its practitioners, and place of practice is that of Renaissance Europe. This form was itself a combination of many ideas from many different intellectual traditions, cultures, and religious contexts, all shoehorned into

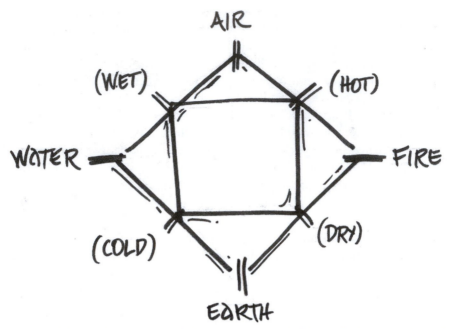

The Greek model of the four humors and elements were part of alchemical thinking and part of medicine for over 1,500 years.

a Christian sensibility, but they were by no means a universal picture.

The practice of something akin to alchemy can be traced back to ancient Babylon. Alchemy in its modern, recognizable form appeared first in the early Moslem period. Following the fall of the Roman Empire (an event commonly seen as the end of the ancient world), two great Western cultural engines appeared—Christianity and Islam—which, in turn, joined with the preexisting Jewish world. For the most part, Christians rejected the learning of the Egyptians, Greeks, and Romans, considering their pagan roots as antithetical to good religious practice. Islamic scholars, however, had few such qualms. In fact, they embraced pagan science and technology as an aid to proper Islamic worship and ritual. The Greek mania for simplifying the universe so it could be better understood led to the creation of the four

elements concept. The Greeks visualized the world around a diagram listing air, fire, earth, and water as the four basic building blocks of the universe. They, in turn, were joined by the conditions of hot, dry, cold, and wet. The entire universe, they believed, was made up of combinations of these four elements and their corresponding conditions. Islamic and later Christian intellectuals embraced this notion as a simple and elegant way to view the workings of the cosmos. The logic behind this arrangement suggested that, if all matter was made of these four simple elements, it would be relatively easy to change one bit of matter into another by simply rearranging the recipe by which they were combined. With that, the popular concept of alchemy was born.

Two of the most famous Islamic alchemists were Al Jabir ibn Hayyan (722–815) and Al Razi (866–925). Jabir, known to his later Christian admirers as Geber, created a number of alchemical apparatus—such as the distillation flask, alembic, and test tubes—that became standard equipment in any alchemical laboratory, and indeed any modern chemical laboratory. He also helped introduce a more disciplined and methodological approach to experimentation and careful laboratory record keeping. The Greek and earlier approach was haphazard and difficult to follow. Despite his improved working technique, Jabir's work was also allegorical and couched in obscure symbolism. The modern word "gibberish" is thought to come from his name. Following the Crusades, alchemy spread into Christian lands where it became a mania in Europe.

The golden age of alchemy was from the late Medieval through Renaissance periods of European history. Renaissance magi came to believe that the originator of alchemy was Hermes Trismigistus (thrice great Hermes), a composite of Greco-Roman deities. As a result, an entire body of knowledge, not all of which was alchemical, was developed based upon the writings attributed to this character and called the *Hermetica*. The first alchemical book to reach Europe was a translation by Englishman Robert of Chester called *The Book of the Composition of Alchemy* (1144). Alchemical books took on a distinctive appearance early. As many were searching for crude ways to turn lead into gold, the alchemists were careful to guard their work against prying eyes. As a result, they wrote their books in increasingly allegorical language and symbolism so that only other initiates could read them.

While the basic underlying idea of alchemy is simple, the practical aspect of reworking the recipes for matter in order to transmute them into other forms proved

Equipment like this furnace and flask, pioneered by alchemists, is still in use by modern chemists.

a huge operational barrier. Practitioners devised a wide range of complex equipment to perform their work. They were primarily concerned with distillation, where liquids were changed to a vapor through boiling and then returned to their liquid state, and with the heating, melting, and cooling of metals and chemicals. The central apparatus in the alchemical laboratory was the furnace. Because different metals melted at different temperatures, there were often several furnaces in the room so that multiple experiments could run simultaneously.

The most obvious impact of alchemy was on the early foundation of chemistry, including metalworking, gunpowder production, dye making, ceramics, glass making, and the development of the alcohol industry. Most modern scientific chemical processes can find their origins in alchemical research. Alchemy also had a role in the growth of modern medicine and the pharmaceutical industry. During the period when disciplines like chemistry and medicine were entering their modern forms, most doctors practiced a level of alchemical research, and some still openly practiced astrology. Possibly the most influential of these alchemist doctors was Paracelsus.

Born Philippus Bombastus Von Hohenheim (1493–1541), he took the name Paracelsus to give himself greater social significance. He was a Swiss physician-astrologer who was also heavily influenced by alchemy. During his youth, he worked as a mining engineer studying rocks, minerals, and strata. Alchemical studies, combined with his practical knowledge of geology, led him to pioneer the use of minerals as cures for medical problems rather than relying solely on astrological, religious, and herbal remedies. He accepted that disease was something brought into the body from outside. Paracelsus saw alchemy not as a way to produce silver and gold to make one rich, but as a means of producing medicines to make one healthy. His approach saw sickness in the alchemical tradition as a disharmony between the individual—the microcosm—and the universe around them—the macrocosm. He argued the human body needed a proper balance of minerals in order to maintain a proper balance of micro and macrocosm, thus good health. His work helped later followers to overturn the Galenic medical philosophy that had dominated the West for centuries.

Alchemy was such an integral part of Medieval and Renaissance thinking that it inspired Isaac Newton to engage in it, though secretly. Scholars are unsure just what Newton's purpose was in taking up alchemy, but it was clearly important to him, as he wrote more on this subject than he did on the straightforward scientific topics such as gravity and optics for which he is best known. The work of Paracelsus and Newton show that, far from being some strange, foreign practice performed by cranks in the dark corners of society, alchemy, in a broad sense, was part of mainstream intellectual thought. It is persuasively argued by historians that alchemical research helped pave the way for later understandings of the universe and was a pivotal intellectual part of the Scientific Revolution, which supposedly did away with superstitious belief for a society based upon reason alone. All of early chemistry took its working methodologies and underlying assumptions directly from alchemy. Chemists eventually dropped the more superstitious and theological aspects of alchemy, but the core structure was retained. It can also be argued that alchemists did not disappear, they simply found better ways of earning a living by turning to rational chemistry. Once dismissed by scientists and historians alike as nothing more than a mildly interesting pseudoscience indulged in by persons of dubious integrity, the modern reappraisal of alchemy, and its resurrection as a worthwhile topic of historical study, came in the late 1970s with the publication of Belgian historian of science Robert Halleux's *Les Textes Alchemique*. He saw the work of some alchemists as organized and experimental and thus forming the basis of modern experimental science. This opened up alchemy as a topic serious scholars could and should investigate. It was this growing body of literature that helped overturn so many of the fantastical and preconceived notions about alchemy.

A modern historiographic difficulty often encountered by those researching and writing about alchemy is its relationship to chemistry. The popular image of the alchemist as a wizard-like character practicing a dark art in a crowded, glass apparatus-filled hovel, attempting to change lead into gold is a misleading one that does not fit the historic facts. The origins of chemistry have been linked to the history of alchemy in popular as well as scholarly texts. The traditional view is that alchemy was a strange, irrational fringe pursuit and that chemistry, as a logical practice, evolved out of it almost accidentally. This view has been repudiated by the scholarship of Lawrence Principe and William Newman. Their close reading and analysis of original texts and primary sources shows that there was no differentiation between alchemy and chemistry to the practitioners of the field prior to about 1700. Principe and Newman suggest using the term *chymistry* to label the interwoven nature of these two historical pursuits. After the early part of the 16th century, however, a division between the two practices did appear, with alchemy veering off into the more spiritual and metaphysical aspect of the endeavor (including the transmutation of metals) and chemistry moving toward its modern form as a non-metaphysical science of materials and atomic structure. As such, *chymistry* was the foundation of the modern pharmaceutical and metallurgical industries and contributed to modern laboratory techniques of a number of sciences. The transmutation aspect can even be seen as an early step toward biological evolution theory.

While many Medieval and Renaissance practitioners of alchemy were frauds, scam artists, and sincere but self-deluded seekers of riches through the transformation of metals, a few were working consciously to turn the irrationality of mysticism into a rational method of understanding the universe. It was this part of the alchemical community that laid the groundwork for the discovery of fundamental ideas about atomic theory, thermodynamics, the periodic table of the elements, and modern science in a wider context.

Alchemists never succeeded in some of their more popular if not common goals. There is little more than anecdotal evidence that any alchemist ever produced a Philosopher's Stone or turned lead into gold. Modern alchemists with all the resources of modern laboratory technology—much of which originated with ancient alchemists—have yet to perform this feat either. As to the other aim of alchemy, the transmutation of one's inner self into something higher and better, that is a metaphysical outcome science can neither prove nor disprove. Alchemy is thus an example of how something that is considered pseudoscience can eventually become science and shows the sometimes transitory nature of intellectual inquiry.

Further Reading

Moran, Bruce T. 2005. *Distilling knowledge: Alchemy, chemistry and the scientific revolution.* Cambridge, MA: Harvard University Press.

Newman, William R. and Lawrence M. Principe. 1998. Alchemy vs. chemistry: The etymological origins of a historiographic mistake, *Early Science and Medicine* 3 (1): 32–65.

ALIEN ABDUCTION

Of all the topics discussed in this book, the most disturbing may be the idea that aliens from other worlds have been coming to earth, kidnapping humans, and performing horrifying experiments upon them for reasons even more repulsive. In North America alone the numbers of people reputedly abducted reaches into the millions. They have been taken from remote areas and urban settings alike. Men report having their semen extracted, while women report being impregnated to give birth to human/alien hybrids. If this is actually happening, alien abduction poses the greatest threat to the human race in its entire history.

When the UFO era began in the late 1940s, the discussion about them centered on what they were and where they come from. By the mid 1950s, that discussion included what kind of occupants rode in them. The first well-known claim of a personal experience with aliens and their flying saucer was made by Polish immigrant George Adamski (1891–1965), who claimed he had been taken aboard such a craft. Adamski was an eccentric amateur philosopher who studied Eastern mysticism and the occult and alternately referred to himself as "Professor" and "Doctor." In 1952, he went into the Mojave Desert of California with some friends to look for UFOs. To their great excitement, a disk appeared and Adamski was invited aboard. Inside he met the ship's captain, Orthon, who told him he was from Venus. The aliens were unusually attractive humanoids who bestowed certain favors upon the bewildered but grateful earthling. In *Flying Saucers Have Landed* (1953) and *Inside the Flying Saucers* (1955), Adamski told of his wondrous adventures with the aliens. It was an optimistic tale full of galactic bonhomie. His books are a mix of fanciful make-believe, light philosophy, and euphoric delusion that, when taken as a body, are rather endearing in their naïveté. He eventually gathered a large following of those eager to become "contactees" and became a kind of guru. Adamski has been alternately praised as a man who changed the world for the better by his contacts with aliens and denounced as a misguided charlatan who made belief in UFOs ridiculous, scaring off the scientific community that, if sufficiently interested, could have solved the mystery. The contactee movement, with its air of hopeful possibility, soon took on a darker aspect. By the 1960s, particularly after the Betty and Barney Hill case, contactees became abductees. Gone were the days of benevolent humanoids in silvery jumpsuits giving lucky humans joyrides in their saucers. Now truly "alien" creatures were snatching up helpless humans in order to perform unspeakable cruelties upon them for sinister ends.

On September 19, 1961 married couple Betty and Barney Hill were returning home from a vacation. As they drove through the White Mountains of New Hampshire, in the United States, near the town of Groveton, they noticed a strange light in the night sky and stopped to look at it. When they arrived home they noticed that several hours of time seemed to have vanished and they had no recollection of what had happened. A troubling and ill-defined sense of unease afflicted both of them for some weeks after. Betty Hill had called the local Pease Air Force Base to report the incident and began reading UFO literature, including works by Donald Keyhoe. They visited doctors for relief but found none. They eventually visited Boston psychiatrist Benjamin Simon in 1963. With Simon, they tried hypnotic regression to see if the missing time could be found. For the next six months, while under hypnosis, the Hills recounted a frightening story of being abducted by aliens who performed disturbing experiments on them. Barney (1923–1969) had a sperm sample taken, while Betty (1919–2004) reported having been pierced by a long needle in what seemed a pregnancy test. Though Dr. Simon did not believe the Hills had actually been kidnapped by aliens, he was convinced the Hills believed they had been, and that something profoundly troubling had happened to them. They spent some time after this in therapy attempting to deal with the trauma. The Hills did not publicize their encounter, but word of it did spread. Research into the case by journalist John Fuller resulted in the book *The Interrupted Journey* (1966), which brought the case to wide notoriety.

There are legions of similar abduction stories. They tend to contain similar elements of missing time, alien entities, frightening experiments, and other details. There is also a long tradition of non-alien abduction stories in human culture where frightening strangers, demons, pixies, witches, and monsters took innocent victims against their will. In North America during the 17th and 18th centuries, European settlers (mostly women) were abducted by Native Americans; sometimes they were quickly released, but other times they were held for years, decades, or even for the remainder of their lives. They would sometimes be integrated into the aboriginal society, with women marrying Indian men and bearing children or men marrying and conceiving children with Indian woman. The popular genre of Indian abduction books became the first big selling literary genre in America. Even more popular were the 19th century "Runaway Nun" books in which helpless and innocent Protestant women and girls were held against their will in nunneries by Catholic, usually Irish or Italian, priests. While they were all eventually proven to be hoaxes, these stories were so troubling and accepted as genuine that they inspired several riots, including the burning of a Catholic nunnery and orphanage outside Boston in 1835.

Alien abduction stories had become so prevalent by the 1990s that a series of large scale investigations were conducted. The first but least well known academic to investigate alien abductions was University of Wyoming psychologist Leo Sprinkle.

He began quietly investigating in the 1960s and became so convinced the phenomena was genuine he even came to believe he himself had been abducted. The most famous academic to investigate the phenomenon was Harvard psychiatrist John Mack (1929–2004). He concluded that the vast majority of abductees were neither deranged nor hoaxers, but sincere people who believed they had been kidnapped by aliens. Another well known abduction investigator is New York artist Bud Hopkins. He saw a UFO in 1964 and began researching the topic. In 1975, his research became widely known when he investigated a series of sightings in New Jersey. Soon he was being contacted, not by aliens, but by people who claimed to have been abducted. He interviewed hundreds of people and he, too, became convinced the experiences were real.

One of the recurring themes in alien abduction narratives is that the victims cannot remember what happened to them; they experience "missing time." Since the Hill case, the most common technique for recovering knowledge of the missing time has been through hypnotherapy. John Mack and Bud Hopkins made extensive use of the technique, which is also known as Past Life Regression Therapy. After being hypnotized by a questioner, the subject is helped to remember past experiences that are difficult or impossible to remember outside of a nonhypnotic state. In the hypnotic state, subjects remember details of their experiences. This technique is highly controversial. Supporters say it is a productive way for trauma survivors to discover the root of their troubles and way overcome them. Detractors argue that the human mind is so complex, and what we call memory so ethereal and easily created, altered, confabulated, and misunderstood that one can never be sure if a recovered memory is genuine. Critics point out that UFO investigators like Bud Hopkins are not trained medical personnel, have at best a superficial understanding of the workings of the brain, and that they can do more harm than good. They can put memories of alien abduction into someone's memory even if they had not been abducted and that, because of the nature of these recovered memories, they cannot be seen as accurate descriptions of events anyway.

Alien abduction proponents have formulated a number of scenarios for why abductions take place. They generally revolve around a need for the aliens to conduct genetic research into human reproduction, which itself is somehow connected to a desire by the aliens to create a human/alien hybrid race. These scenarios tend not to have happy endings. There have been a number of cases in which abductees claim to have been implanted with alien-designed "tracking devices." Such implants, it is argued, allow victims to be abducted multiple times over the course of their lives. Odd bits of metallic materials have been removed from abductees and analyzed, but none have proven to be otherworldly or of any recognizable technology. Dismissed by mainstream science and medicine as seekers of fame, sincere but deluded people, or simply mentally disturbed or distraught, abductees have formed their own subculture and support groups where they can exchange information without the fear of ridicule and help comfort each other.

See also: Past Life Regression; Unidentified Flying Objects (UFOs).

Further Reading

Clark, Jerome. 1998. *The UFO book: Encyclopedia of the extraterrestrial.* Canton, MI: Visible Ink.

Jacobs, David M. 1993. *Secret life: Firsthand, documented accounts of UFO abductions.* New York: Touchstone.

Mack, John. 1995. *Abduction: Human encounters with aliens.* New York: Ballantine Books.

ALIEN AUTOPSY

Film footage released to the public in August 1995 purported to show government operatives performing an autopsy—a postmortem examination—on the body of a being from another world. The film was promoted by British film producer Ray Santilli. He claimed the film was shot in 1947, just after the infamous Roswell Incident. According to ufologists, an alien spacecraft crash landed in the New Mexico desert near the town of Roswell and was recovered by the U.S. military. Along with the craft itself, several crewmembers' bodies were recovered, and this was film of the autopsy of one of them. The film was immediately controversial and was shown worldwide as a documentary hosted by Jonathan Frakes, the actor who portrayed First Officer Will Ryker on *Star Trek: The Next Generation.*

Santilli claimed he had discovered the footage in Cleveland, Ohio in 1993 while searching for vintage film footage of 1950s rock-and-roll stars. It was said to have been part of a cache of 16mm footage shot by a U.S. Navy cameraman, but which had become separated from other reels and forgotten. The film was left to sit in a box, along with a few other reels, for years, until Santilli happened along. In the grainy, black and white, silent film, a group of individuals wearing protective suits can be seen standing in an operating room examining and dissecting a large headed, goggle-eyed, humanoid. They make various incisions and remove organs from the being, which has many obvious wounds.

Santilli steadfastly refused to divulge details of the film's origins or acquisition. A number of Hollywood special effects experts reportedly claimed it was so well done they could not figure out how it was made. They also claimed it would have cost hundreds or thousands of dollars to make. In reality, most special effects professionals had said the film looked fake and obviously so. The Committee for the Scientific Investigation of Claims of the Paranormal (CSICOP) was also was quick to denounce the film as a hoax. A brace of other autopsy films of varying quality followed in the original's wake, as films and television shows, some produced by amateur filmmakers, showed how easy it was to create such a piece. In a 2006 interview, Santilli confessed to hoaxing the entire thing, sort of. He claimed that the film was a restoration of the original. He said that he had viewed the original, taken by the Navy photographer, but it was mysteriously lost before he could buy it. Santilli then had a reconstruction made and the few remaining bits of the original spliced into it. The filmed "interview" of the Navy photographer was also staged. The interviewer on that bit of footage was Robert Kiviat, who was later involved in the debunking of Patterson Film, which claimed to show Bigfoot and appeared as part of Greg Long's *The Making of Bigfoot* (2007).

See also: Alien Abductions; Unidentified Flying Objects (UFOs).

ALTERNATIVE MEDICINE

A broad range of healing methods practiced outside the mainstream medical community. A diverse range of therapeutic practices fall into this category because they are considered unconventional, lack biomedical explanations and definitive scientific proof, or are simply not taught or practiced in traditional Western medical schools and hospitals. Treatments tend toward the holistic: treating a person's whole body, as well as their environment. The focus is typically on attaining spiritual and physical balance and addressing obvious physical symptoms as well as issues relating to the "inner self." The alternative medicine approach concentrates on the root of an issue, rather than merely its symptoms, as in conventional allopathic medicine. The term "allopathic" was coined in 1842 by C.F.S. Hahnemann (1755–1843), a popular "irregular" medical practitioner, to separate homeopathy as he practiced it from the work of "regular" or professional doctors. Remedies are often naturalistic or derived from natural sources, are gentler, and are claimed to have fewer detrimental side effects than synthetic medicines. Although mainstream doctors sometimes use the term "alternative medicine" as a pejorative, practitioners who are self-professed alternative to orthodox treatments also embrace and use it.

A wide variety of techniques and approaches fall under the umbrella of alternative medicine. Included are the healing properties of color therapy, homeopathy (a complete system of healing that employs natural substances to stimulate the body's natural defenses and aid in recovery), iridology (a diagnostic method used to determine disease), and polarity therapy (an inclusive health system of bodywork, diet, exercise, and counseling). Therapeutic plants and herbs are often implemented in healing and have been used since ancient times. Natural plant preparations like those used in herbal medicine or flower essence therapy are said to support the body's natural healing processes. In liquid form, these plant essences are said to restore balance and good health to the interwoven mind and body. Other naturopathic alternative medicines focus on a person's posture and mobility. For instance, the Alexander Technique is a series of physiological movements designed to bring about conscious awareness, therapeutic relief, and better "posture and mobility." Manipulation therapies like chiropractic, osteopathy, cranial osteopathy, and Rolfing (once known as structural integration) also increase mobility by correcting misalignments in the body. Psychotherapy and its many subdivisions fall under this category in healing maladies of the mind. Once considered bizarre and abstract, the "talking cure" of Sigmund Freud has now not only become commonplace, but it has also altered traditional ways of thinking about the mind and launched a revolutionary industry dedicated to mental health. Hypnotherapy, which was also once feared or considered quackery, has been implemented as a helpful tool in correcting imbalances in a person's subconscious behavior, too. There are many other treatments that center on the mind to balance physical, mental, and emotional levels, and may also include vitality and spirit. Meditation is thought to have soothing beneficial effects. In a similar fashion, autogenic training, visualization therapy, and music therapy are also designed to relieve stress and heal the body. To promote stress relief and relaxation, massage therapy in widely accepted as legitimately beneficial to sore muscles. Some practitioners also believe massage improves overall health and prevents disease, especially when used in combination with other modalities such as reflexology, which imple-

ments the human touch. Aromatherapy is often accessorized with these techniques.

Many of the alternative medicine therapies used today originated in the East. With its supposed reliance on naturopathic remedies, Eastern medicine came into vogue in the West during the 20th century. Patients disillusioned with Western approaches have flocked to anything labeled "Eastern Medicine." Unlike the West's medical culture, which tends to focus on isolated areas of disease and symptoms, Eastern medicine sees the interaction between body, mind, spirit, and relationship with the environment as essential to healing and health. These components are said to encourage well-being and prevent disease, as well as diagnose and treat. Some commonly applied treatments include acupuncture, acupressure, *T'ai chi ch'uan,* herbal medicine from China, and Shiatsu (finger pressure massage) from Japan.

There are a number of diagnostic techniques utilized in alternative medicine. Iridology, for instance, considers the eyes to be a map of the body. Examining the irises is said to be a way of pinpointing health weaknesses. Kinesiology is a series of muscle tests that evaluate body health, and biofeedback monitors responses to various physical tests through the use of machines. Kirlian photography is a technique that is supposed to be able to photograph the quality of a person's energy, which is said to appear altered if there is an imbalance. In reflexology, the sensitivity to finger pressure on a person's feet is indicative of imbalances as well. A health evaluation using Chinese medicine takes the whole body into account. The tongue is examined, the pulse is taken, the body is inspected and palpated, and only then can an assessment be made.

Some alternative medicines are thought to have been in practice for thousands of years, and the regions from which they originate vary greatly. The oldest practices still available today are aromatherapy, which began in ancient Egypt and was expanded upon by the Greek Roman cultures, and Chinese Medicine, which includes acupuncture and Eastern herbalism. Both have been in use since about 3000 BC. Western herbalism, a variation of the formulas utilized in the East, was developed by Egyptians, Greeks, and later the Romans; it can be traced as far back as 2300 BC. By 1000 BC, Ayurvedic medicine had been established in India and was expanded upon by Hindu philosophy.

The Greek father of medicine, Hippocrates, was one of the earliest designers of nutritional therapy. Around 370 BC, he created a form of nutritional therapy that he based on earlier herbal healing traditions. Centuries later, cures continued to be cultivated based upon his early teachings. German doctor Samuel Hahnemann (1755–1843) first conceived of the notion of homeopathy late in the 1700s. In that same time period, Austrian Anton Mesmer (1734–1815) introduced the idea of hypnotherapy. Naturopath Vincent Priessnitz (1799–1851), who also resided within the Austrian Empire founded, "the nature cure" and popularized the system of hydrotherapy as well. Appalled by the surgical methods used in the Civil War, Andrew Taylor Still devised the system of osteopathy in the United States in 1874. This was quickly followed with the chiropractic technique, introduced in 1895. Reflexology emerged at the turn of the 20th century, based on medical principles from the Far East. Australian actor turned physiotherapist Frederick Alexander (1869–1955) developed his Alexander Technique of proper posture, and in the 1930s, Dr. Edward Bach of the United Kingdom introduced the Bach Flower Remedies. This technique was designed to treat negative emotions with the balancing effect of flowers.

One concern over alternative medicine is that many forms are commonly used yet fail to show any definitive proof of efficacy. Some are intentional scams meant to fool patients. Operating outside of the conventional standards may allow some ideas to form and flourish without restriction, but it also invites an opportunity for fraud. Unproven remedies may rely heavily on the placebo effect and marketing to sell products. Those excluded from the physician's elite certification may argue that doctors assert an omniscient status for themselves in order to eliminate competition or for financial gain. In opposition to this view, some standards must be in place to keep the public safe from the distribution of inert or harmful products. Some reasons patrons give for choosing alternative medicine over orthodox healthcare are distrust of an enigmatic healthcare system, capitalist drug companies' impossibly expensive drugs with hazardous side effects, the limited availability of doctors in their area, and limited current medical knowledge about some conditions. However, the realm of unregulated treatment has a sordid history of victims falling prey to the wiles of "snake-oil salesmen" touting "cure-alls" at "medicine shows." It is the undiscerning charlatans or quacks who are cause for concern when stepping outside the conventional medical modalities. While fruitless folk medicines, faith healers, and diet fads tend to fade once proven ineffective, the unfettered alternative medicine term can unfortunately be used as a guise for another ineffective product.

In 1976, the term "complementary medicine" was coined to define the integration of alternative and traditional medical treatments, as some mainstream doctors came to recognize the efficacy of some alternative approaches. Since then, organizations like the American College for Advancement in Medicine (FAIM), the Untied States' largest professional organization of complementary-oriented physicians, have tried to bridge the gap between the holistic and conventional philosophies and techniques. FAIM has also assisted in pushing alternative medicine into the mainstream when they helped to coordinate New York state's first law specifically authorizing the use of complementary and alternative therapies. The American Association for Health Freedom (AAHF) is another group of health professionals that provides an influential presence for complementary medicine on the national level. The Access to Medical Treatment Act (first introduced to the U.S. government in the late 1990s) is a bill sponsored by several members of Congress who hope to transform the practice of complementary and alternative medicine in the United States.

See also: Acupuncture; Homeopathy.

Further Reading

Bivins, Roberta. 2007. *Alternative medicine?: A history*. Oxford: Oxford University Press.
Mayo Clinic book of alternative medicine: The new approach to using the best of natural therapies and conventional medicine. 2007. New York: Time Inc. Home Entertainment.
Tovey, Philip. 2004. *The mainstreaming of complementary and alternative medicine: Studies in social context.* New York: Routledge.

ANCIENT ASTRONAUT THEORY

This theory proposes that various extraterrestrial visitors came to earth in the distant past and somehow ignited or advanced human evolution and civilization. More a collection of presuppositions and presumptions rather than a single coherent body of thought, ancient astronaut theory (AAT) claims that sentient beings from other worlds came to the earth and left their mark on both the physical landscape and the human condition. Variations on the theme include aliens coming to earth intentionally for some as yet unknown reason, or accidentally through some mechanical failure of their technology and becoming marooned here. Once here, the aliens intentionally altered protohuman genetic material as part of a test, mated with early hominids to create *Homo sapiens,* or they mated with and carefully guided the already existing *Homo sapiens* population. All evidence for AAT is speculative and circumstantial at best.

Evidence for ancient astronauts is based upon similarities in widespread human culture or the supposed inexplicability of certain societal advancements and architectural achievements, advances beyond what, according to proponents, such civilizations should not have been capable of. Other rationale for belief in AAT include religious or mythological stories of phenomenal gods or visitors and knowledge of the Earth which could only be attained by traveling extreme heights or distances, not a plausible feat for primitive people. The seemingly rapid explosion of civilizations is also cited as possible evidence for the theory, as well as the evolutionary jump from ape to man. Major proponents of the idea include Charles Hoy Fort and his *Book of the Damned* (1919), and Erich von Däniken and his *Chariots of the Gods* (1968). While Fort's book garnered interest in the subject, it was Von Däniken's that captured the popular imagination and helped spawn an enormous wave of interest in the subject.

In addition to the idea that sentient beings came to earth, there is the less sensational idea that alien microbes may have arrived on earth after drifting through space. Some theories propose that ancient astronauts are responsible for bringing all forms of life to earth, beginning with spores. Distributing or planting primeval spores on the earth in its early stages may have been all that was needed to trigger an eruption of life forms. The purpose of cultivating plant and animal life could have been for scientific experimentation, to create a habitable residence for themselves, or possibly as a hedge against future necessity, when he alien civilization might need the earth for colonization purposes.

The supposed inability of scientists to provide the missing link in man's evolution leads into the idea that visitors may have interbred with the early hominids. If they had colonized during this stage in man's development, it would have been before the last ice age, and remnants of their stay would have been destroyed. Therefore, this idea is made plausible, not by an artifact, but by the leap in development of Neanderthals and Cro-Magnons; both species exhibited a cranium capacity one and a half times larger than primeval man, and 200 cubic centimeters larger than today's average. Alfred Russel Wallace, a contemporary of Charles Darwin, broached the question of whether or not something inexplicable triggered this brain growth in humans, later known as the "Brain's Big Bang."

Other ideas about ancient astronauts are linked to the advancement of civilization. Mysterious ancient temples, which seem unlikely given the time in which they were built, or structures seen only from the sky, are often attributed to otherworldly aides. Cross-cultural similarities in monument construction, despite vast distance and supposed unfamiliarity, fall into this category as well.

One popular subject for AAT theorists can be found in the lines drawn in the surface of the earth at Nazca, Peru. The Lines of the Nazca, a favorite topic of Von Däniken, were formed at least 2,000 years ago by local people scraping the surface of the ground to expose lighter earth underneath. Thousands of lines were drawn. Many were simply straight lines, while others form shapes of animals and insects. Others demarcate the summer solstice line as well as points of the compass, sunrise, and sunset. It is argued that the lines are visible only from the air and some believe them to have been created for religious purposes, although it is unknown exactly why they were created. There are similar unexplained carvings in the Titus Canyon in Death Valley, and the Mojave Maze in California, in the United States. Ancient astronaut enthusiasts credit the formation of these lines as evidence of aeronautic assistance.

In Nazca, some lines point toward a pre-Incan temple in Tiahanaco. It is estimated that 100,000 people were needed to build this "Gate of the Sun," yet there is no evidence showing that the plateau it sits on, 13,000 feet above sea level, was ever able to support the crops or livestock needed to support such manpower. The precision in the construction of the temple walls has kept it intact, and it is incredibly similar to the fittings of the submerged limestone Bimini Wall in the Bahamas. There is some speculation as to whether this underwater monument was perhaps the fabled lost city of Atlantis, which was said to be an advanced civilization even by today's standards. Another curious feature about the Peruvian "Gate of the Sun" structure is the "Gallery of

Early people left drawings like these on rock walls and in caves around the world. Modern researchers claim they are drawings of ancient astronauts who visited the earth.

Mankind," which showcases a variety of dissimilar physiognomic faces. It is unknown how this isolated, primitive culture knew the different races of people across the world.

The appearances of pyramids in Egypt and throughout Mesoamerica are thought to be connected because of their likeness and grandeur. The Great Pyramid of Giza entices pseudoscience enthusiasts because of its alignment with true north on one side and the Earth's polar axis on its east face. Other large structures include Stonehenge and the huge Moai megaliths of Easter Island. Even larger are the three megalithic stones in Baalbek, Lebanon. They are known as the "Trilithon" and are the largest stones to have been quarried, at an estimated 500 tons each, a weight seemingly beyond the capabilities of ancient man.

There are many more artifacts and instances attributed to ancient astronauts. Pre-Incan gold carvings discovered in Colombia resemble spacemen and delta wing jetfighters. While some argue that the airplane carvings are zoomorphic, a vertical stabilizer is featured placing it in a category representative of modern aircraft as opposed to any known bird or insect. A model airplane was also discovered in a tomb in Saqquara, Egypt in 1898. Classified as a "wooden bird model" and shelved in a museum basement when found, the models were later identified as clear flying devices and dated to 200 BC. Other classical Egyptian wall decorations are said to include the outline of an airplane, a helicopter, and an electric light bulb. Egyptologists who have studied the hieroglyphs extensively say this is nonsense.

Vimanas, flying machines, are also mentioned and mapped out in the Sanskrit epics of India. The *Vaimanika Sastra* is an eight-chapter diagramed description of three types of aircraft, including apparatuses that could neither catch on fire nor break. It mentions 31 essential parts of these vehicles and 16 materials from which they are constructed. The vehicles can absorb light and heat; for this reason, they were considered suitable for the construction of *Vimanas*.

Artistic evidence in favor of AAT includes Paleolithic cave paintings from Val Camonica, Italy, dating to 10,000 BC, and at Vondijina, in Australia. The dome-shaped heads of the paintings' subjects have been claimed to depict extraterrestrial visitors wearing what look like space helmets, rather than the mythological persons or gods of archaeological scholarship. Another speculation is the *Dogū*, which is said to be an ancient astronaut who visited the Earth during the Jōmon period of Ancient Japan and resembles a being in a space suit, goggles, and a space helmet. Medieval and Renaissance art are used to support this theory as well. It is suggested that certain objects in the paintings that cannot be easily explained with relevance may indeed be flying saucers. These objects are used to support AAT by attempting to show that the creators of humanity return to check up on their creation throughout time.

Various religious creation stories begin with two or more beings coming to earth. There are Chinese theories about an alien race called the *Dropa* who left behind fascinating discs. The Mayan religious text, the *Popol Vuh,* clearly states, "Men came from the stars, knowing everything, and they examined the four corners of the sky and the Earth's round surface."

In an example of how AAT has excited the interest of engineers, Joseph Blumrich of NASA took special note of the Bible story of Ezekiel after seeing Von Däniken's discussion of it. He decided to try to recreate the image as described in the Bible. He built a scale model flying machine based on the story. Ezekiel, in his first vision, saw a great cloud with fire flashing through it and radiance around it, something that gleamed like metal and the sound of great waters. Blumrich likened this to the downdraft and tail of jet exhaust during the descent of our lunar landing crafts. In verse seven, the description of straight legs with hoofs like calves glittering like polished brass parallels the aerospace structure determined to be the most logical shape for an unmanned landing craft, as devised by Blumrich and NASA between 1962 and 1964. Ezekiel's account of four figures, each with four wings, that descended through the clouds, and one of which lifted him up between the earth and heavens to the gate of Jerusalem and back to Chebar in chapter 8 can be equated to the design of helicopters, as Blumrich was able to show in sketches. Similar descriptions can be found in several other reports by Ezekiel. Other biblical figures like Elijah, who departed earth in a chariot of fire, and Enoch, who walked with God and was taken up by a whirlwind, have also been reevaluated as possible encounters as well. The explanation for these somewhat confusing accounts centers on the fact that there was no vernacular at the time to describe spacecrafts accurately to our modern understanding, but that the prophet was describing an actual alien machine. Blumrich explained his theory in *The Spaceships of Ezekiel* (1974) and inspired many to examine other biblical accounts for traces of alien visitation.

While the idea of ancient alien space travelers coming to the earth to jump-start human civilization is a romantic and tantalizing one, it has its dark side. The theory suggests that all the great human accomplishments of the ancient world were not our own, that ancient people, often dark-skinned ancient people, were mentally incapable of achieving complex feats of science and engineering. AAT relegates humans to background characters, groveling in the mud like simpletons who need alien super-beings to come save them and even build their shelters. Scholarly studies of these civilizations shows that ancient humans were more than capable of building pyramids and other monumental architecture on their own. They created complex societies and accomplished astounding feats without the help of anti-gravity device-toting and flying saucer-riding aliens. In short, the AAT is a modern recasting of 18th- and 19th-century European attitudes about "colonials" and "natives."

See also: Fort, Charles Hoy; Von Däniken, Eric.

Further Reading

Blumrich, Joseph. 1974. *The Spaceships of Eze-kiel*. New York: Bantam.

Fort, Charles. 1919. *Book of the Damned*. New York: Boni & Liveright.

Von Däniken, Eric. 1970. *Chariots of the Gods*. New York: G.P. Putnam's Sons.

ANIMAL TESTING

The use of animals as subjects to generate scientific and medical evidence through experimentation. Wide ranges of animals are used for experiments, including mammals (rodents, primates, cats, dogs, and rabbits mostly) and invertebrates (insects), fish, and occasionally birds. Advocates of animal testing argue that the practice is good science that produces data crucial for developing medicines and combating disease and to help learn the effects of commercial and industrial products upon humans. Detractors claim it is pseudoscience, as the tests cause undue harm and suffering to the animals, the results generated are of dubious use for human research, and that better results can be generated through non-animal based alternatives.

The use of animals for testing goes back to the ancient world. Greek philosophers, including Aristotle, used animals in their work. The Roman physician and father of modern medicine, Galen (AD 129–200), was prohibited by social custom from dissecting humans, so he used animals to learn human anatomy by analogy. By the 19th century, animal experimentation reached a level where it began to cause concern among ethicists and animal enthusiasts. Animal experimentation and dissection, called vivisection (to cut up), was done on live animals as well as dead ones. Public knowledge and outrage over these practices (anti-vivisection) became the first organized resistance to animal testing. The anti-vivisection movement spread across Europe, the United Kingdom, and North America. Its effects are debatable, as the scientific and medical community had powerful government backing as well as popular support. It was easy to argue in favor of animal testing, as the medical community could point to successes like Louis Pasteur's use of animals to work out germ theory and a rabies vaccination, to name just one example. Animal testing opponents increasingly sought to connect themselves to the separate, but more popular anti-animal cruelty movement. The modern animal rights movement finds its genesis in this combination.

By the middle of the 19th century, organizations like the American Society for the Prevention of Cruelty to Animals (ASPCA) and similar organizations in the United Kingdom and Europe helped spread the popularity of the anti-cruelty and anti-animal testing movements and helped pressure governments to begin passing related laws. By the 1960s, the American Laboratory Animal Welfare Act was passed. Throughout the century, governments worldwide, under pressure for the public, passed an increasing number of animal welfare laws and experimentation regulations. These laws included specifics about how animals were to be housed, fed, generally treated, and how they were to be destroyed, or euthanized, once their jobs were complete (animal rights activists prefer to say "killed").

Amongst anti-animal testing advocates there are a number of intellectual approaches to the problem. The pro-animal, or animal rights, school argues the ethics of using animals in painful and often lethal testing. They focus on the idea that animals have intrinsic rights (as humans do) and therefore should not be experimented upon any more than a human should be. Animals, they argue, have their own right to life regardless of how humans feel. They also argue that no matter how humane the animal's treatment is, the very notion that anything useful about human anatomy can be learned from animals about human health is pseudoscientific and, as such, should be abandoned for more useful, non-animal centered approaches. Animal testing, they argue produces results that are not only useless, but also could be misleading and thus dangerous to humans. They point to alternative testing methods, particularly computer modeling, which not does not harm animals and produces efficacious results.

The most public and controversial type of animal testing is that which is related to toxicology. This safety testing method uses animals to measure or determine the harmful side effects of food additives, shampoos, hair dyes, and cosmetics. Such procedures, like the Draize test, involve putting toxic elements into the eyes of cats and rabbits. This produces burns, irritation, and other painful results. After years of pressure from animal rights groups, a number of European countries and the United Kingdom began to pass laws banning cosmetics testing (though not all testing) on animals. Pharmaceutical companies continue to fight such regulations. Countries like Belgium and New Zealand have passed laws prohibiting experimentation on higher primates such as chimpanzees and gorillas. Some anti-animal testing activists argue that even these laws are ultimately wrongheaded, asserting that all animals have moral rights, and so their species should be irrelevant.

The most well known animal's rights organization is People for the Ethical Treatment of Animals (PETA). They have raised awareness of the issue with deftly managed advertising campaigns, propaganda, and efforts to get legislation passed. Undercover "whistleblowers" and journalistic exposés have shown conditions in some laboratories and corporate offices where animal testing is done to be places where workers routinely practice animal cruelty. Hitting, isolation, humiliation, starvation, and other cruelties have been documented well beyond the testing itself. As a result, firms have been fined and embarrassed and workers fined and fired.

Some animal rights advocates, including some PETA members, have abandoned lawful ways of addressing the issue and moved to violence to achieve their ends. The most notorious group is the Animal Liberation Front (ALF), which seeks to end animal testing by hurting companies and universities that engage in actions they deem unacceptable by affecting them financially. As the ALF Web site says, "usually through damage and destruction of property." They have broken into laboratories to rescue animals and damage the facilities, and claim they take the rescued animals to sanctuaries. While they also claim they do not want to hurt humans in their actions, ALF has been linked to a number of incidents in which researchers and their families have been targeted for harassment and intimidation, including threats of violence. ALF's motivations may be honorable, their actions, place them in the same league as radical anti-abortionists who seek to end that practice by blowing up health clinics and murdering doctors. ALF has yet to kill a scientist, but it is feared that it is only a matter of time. In January of 2009, ALF members proudly proclaimed they had sent "letter bombs" to researchers at the University of California at Davis, a facility that does research focused on AIDS and Alzheimer's disease.

Animal testing brings up questions of ethics as well as science. Done in laboratory settings with the trappings of science, detractors (including many scientists) say animal testing is pseudoscience because it simply does not produce results. If this is the case, then we are confronted with the question of what to do with pseudoscience, especially if it is harmful. Should we legislate it away and possibly impede other scientific research that does improve the human condition, or should we do as ALF wants and blow it all up? It is generally thought that even the highest cognitively functioning mammals, like higher primates and dolphins, do not think or feel the way humans do. Does that matter? What if, like many subjects once thought to be pseudoscience, we find that these animals do indeed think the way we do? After all the years of animal testing, how would we come to think of ourselves?

Further Reading

French, Richard. 1975. *Anti vivisection and medical science in Victorian society*. Princeton, NJ: Princeton University Press.

Greek, Jean Swingle and C. Ray Greek. 2006. *What will we do if we don't experiment on animals? Medical research for the twenty-first century*. Victoria, BC: Trafford Publishing.

Guerrini, Anita. 2003. *Experimenting with humans and animals: from Galen to Animal Rights*. Baltimore, MD: Johns Hopkins University Press.

ANOMALOUS BIG CATS

Also called Alien Big Cats, these are large felines that appear in places they do not normally inhabit. Most commonly associated with the British Isles, but also seen in the Americas and occasionally Europe, Anomalous Big Cats (ABCs) are out of place tigers, pumas, mountain lions, and similar creatures seen roaming wild areas where they are not indigenous. Unlike their cousins the phantom dogs, ABCs are not usually associated with omens or catastrophes, nor are they normally seen as spectral entities. Illegally released pets or zoo and circus escapees account for some of these sightings, but the entire number of sightings cannot be accounted for by these explanations alone.

Like North America's anomalous primates, the ABCs of the British Isles excite interest, appear in rural areas, make only brief appearances, and appear in only a few blurry and indistinct photos and films. The few films of ABCs taken in the United States look suspiciously like house cats exploring farmers' fields. Studies by local wildlife authorities invariably end inconclusively. When a large cat skull was found in the River Fowey and brought to

Anomalous Big Cats like the Beast of Bodmin are difficult to track or catch and in photos often look suspiciously like house cats.

the British Museum of Natural History, an examination showed it to be a skull cut from a leopard skin rug.

Like anomalous primates, ABCs have their star performers. Two of the best known ABCs are the Beast of Exmoor and the Beast of Bodmin. The Exmoor cat, of Devon, England, was first reported in 1970. It gained an infamous reputation in 1983 when it was accused of single-handedly (or single pawed) slaughtering a hundred sheep on a single farm over a period of just three months. It is said to resemble a black panther in size and color. A similar creature is the Beast of Bodmin, reported in Cornwall. All the ABCs tend to perform in the same way.

Big cats are appropriate for the British Isles. They are just the right scale for the wild places in which they hide. One of the reasons Bigfoot-like creatures are not found in the region is the problem of where to hide an eight-foot-tall, 500-pound bipedal primate in the Devonshire countryside. Anomalous primates have plenty of room to hide in North America or the frozen wastes of central Asia. Big

cats do inhabit the Americas, and do sometimes even wander into human-inhabited areas (when this happens, it usually ends badly, with the cat attacking a human or the cat being dispatched by armed citizens or the police). Cats, even large ones, can conceivably hide in the British countryside, lurking in the brush living off mice and squirrels and going undetected.

See also: Cryptozoology.

Further Reading

Beer, Trevor. 1986. *The beast of Exmoor: Fact or legend?* UK: Countryside Productions.

Campion-Vincent, Veronique. 1992. Appearance of beasts and mystery cats in France," *Folklore* 103 (2): 160–83.

ANOMALOUS FOSSILS

Fossils and fossil-like objects that turn up in forms or locations conventional wisdom says they should not. For example, fossils of animals known to have existed at a certain point in geologic time but are found earlier than that; fossils known to be found in certain geographic locations that are found elsewhere; mixes of incongruous fossils or modern artifacts found as fossils. Mainstream science acknowledges the existence of most anomalous fossils and has prosaic ways of explaining them. There is another class of anomalous fossils that are rejected by science but championed by creationists and other anti-evolutionists as proof that evolution does not work. It is this class of anomalous fossils that constitute pseudoscience. Each of these examples has turned out to be a misidentification or hoax.

The word "fossil" comes from the Latin *fossus* for something dug out of the ground. There are two types of genuine fossils: the remains of once-living creatures turned to stone or otherwise preserved, and fossils made by the actions of living things—footprints, impressions of sea shells—or nature—raindrops, waves on a beach, and the like. There are a number of ways the remains of a living thing can become a fossil, but the most common is for the remains to be saturated and replaced by sediment which eventually turns to stone, eliminating the original organic material but leaving an exact copy. It takes anywhere from 10,000 to 15,000 years for this process to take place. The ratio of fossils compared to the original population of organisms is low; only tiny fractions of all organisms ever become fossils. The age of fossils is determined through radiometric dating, which measures the atomic decay rate of the surrounding stone, or matrix, in which the fossil is lodged. Radiometric dating has advanced to a point that it can give highly accurate and reliable results.

Creationists often point to geologic formations and conventional fossils to argue that, far from supporting evolution, they actually support the creation model. They also employ the most dubious examples. Creationists have searched the world and written records for peculiar fossils, ones that would topple and upset accepted notions about the course of human evolution. One of the most often cited of these anomalous fossils is the "Meister Print," from Utah. This strange artifact was found in 1968 by fossil hunter William Meister. He was prospecting for trilobites in 500 million-year-old strata known as the Wheeler Formation when he cracked open a slab to find what looked like a human shoe print that had crushed a live trilobite under its heel. Anti-evolutionists and anomalists seized upon the print as evidence that humans had trod the earth well before the time scientists said they did. The Meister Print was held as proof that the fossil record, the geologic time scale, and the very notion of human evolution were false. Several studies showed the print was, in reality, an example of a common geologic occurrence known as spalling, in which slabs of rock break away from each other in distinctive patterns. This particular case of spalling had created a simulacrum vaguely suggestive of a shoe print. After this determination of prosaic natural causes was put forth in the 1980s, some creationists refused to acknowledge the print, while others continued to recognize it.

Similar fossil footprints have turned up fairly regularly since the 19th century, and each time creationists got excited that this may be the one that would destroy the edifice of evolutionary thinking. There have been supposed human skulls found under lava millions of years old; in the Americas, coins were found embedded in stone from ages before coins were first manufactured; and, one of the most popular creationist artifacts, a modern hammer has been found locked in a mass of sedimentary rock. These discoveries were gleefully put forward with calls for scientists to explain them. When examined, they all turned out to be misidentifications of other things, hoaxes, or, like the Meister Print, simply examples of people not understanding what they are looking at.

A major compendium of such anomalous fossils and other evidence of the falsity of human evolution theory can be found in *Forbidden Archaeology* (1994) by Michael Cremo and Richard Thompson. Besides the sheer mass of material they collected, what makes this book unique within the genre is that the authors are not Christian fundamentalists, but rather Hindu followers of Krishna Consciousness. Although they cited much the same material as the Christians and other antievolutionists, they argued the human race was not very young but instead immensely old—far older than even evolutionists are willing to allow. They claimed modern humans appeared through divine intervention more than 55 million years ago and that ancient Vedic scripture told the true story of how humans came to be. That their thesis was different from other antievolutionists did not keep them from falling into the same traps. They quoted evolutionists out of context and accepted vague, century-old accounts of anomalous fossils while dismissing modern scientific ones backed up with meticulously gathered documentation. This is a common technique in antievolution and anomalist circles. Cremo and Thompson argue that there is a concerted effort on the part of the "evolution establishment" to suppress any evidence that contradicts mainstream thinking, especially mainstream human evolutionary thinking. They share this notion with others. In their sem-

inal work on anomalous artifacts and events, *Phenomena: A Book of Wonders* (1979), John Michell and Bob Rickard argued that "some of these evidences have been actually suppressed and deliberately ignored: some have quietly and mysteriously disappeared from museum stores and records" (75). Yet they gave the devil his due by saying that accepting that humans were around hundreds of millions of years ago based on such scanty evidence was "irrational." Since the 1990s, Christian creationists have claimed that the Smithsonian Institution in Washington, DC has housed a collection of artifacts that were obviously intelligently designed, but that the designers were and what purposes the artifacts served were unknown. The Smithsonian disavows having any such collection.

An enduring belief within creationist circles has been that humans coexisted with the dinosaurs, and if proof could be found that they did the entire human evolution edifice would crumble. To support the contention, believers point to the "man tracks" of Texas. They argue that what paleontologists call dinosaur tracks are, in reality, human footprints frozen forever in mud and turned to stone. The tracks were first discovered in 1917 along the banks of the Paluxy River outside the town of Glen Rose, Texas. That region of Texas had been a floodplain 70 million years ago, as the Gulf of Mexico extended much farther inland than it does today. Many of the tracks were clearly those of dinosaurs, but a few took on odd shapes because of the distortion caused by the animals slipping and sliding in the wet mud. It was these tracks that creationists argued were human footprints. Roland T. Bird (1899–1978), the American Museum of Natural History excavator who did the first extensive work at

Glen Rose in the 1930s, tried to convince people that the prints were just deformed dinosaur tracks.

Two recent anomalous fossils that have become the rage in antievolutionist and creationist circles are the "Limestone Cowboy boot," and the "clock in the rock." The former is an obviously modern cowboy boot that is encrusted with a stony matrix in which are encased a number of bones. Creationists Carl Baugh and Donald Patton have been the primary promoters of the cowboy boot, which is said to have been discovered in 1980 near the Texas town of Iraan. The boot was determined to be of a type that could be accurately dated to 1950. A number of creationists and Young Earth proponents have also claimed the boot is "40 Million Years Old!" No tests have been done to authenticate the boot's fossil status. Tests have been made, but the results have not been made public as of yet. Photos of the artifact even bring in to doubt if the bones inside the boot are human. Baugh has since stepped away from claims about the boot. Patton has a history of promoting the modernity of "fossil humans" that have turned out to be recent burials, Malachite Man being one well-known example. At the 2005 Creation Mega Conference in Virginia, Patton famously stated that one need not know anything about human anatomy in order to know human evolution was a sham.

The "clock in the rock" is a slightly older object that was found near Westport, Washington in 1975. It is the remains of the internal parts of a wind-up clock encrusted with a hard matrix including seashells. Geologists pointed out that there are certain materials, like beach sand and dirt, which can, under the proper conditions, harden quickly around an object giving it the superficial appearance of a

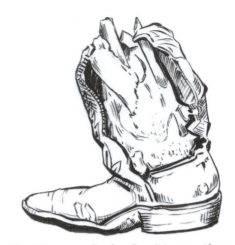

The "Limestone Cowboy Boot" is an artifact creationists think proves evolution does not exist and that the earth is really very young.

fossil, but that these are not true examples of fossils. Having apparently learned not to make such claims, well known Creationist entrepreneur and head of the group Answers In Genesis (AIG) Ken Ham, instead of arguing it represented a modern artifact in ancient stone, said that the clock was an example of how it does not take very long to make a fossil. This allows him to claim that there are no real fossils of the ages scientists claim, therefore evolution is false.

See also: Creation Science.

Further Reading

Bird, Roland T. 1985. *Bones for Barnum Brown*. Fort Worth, TX: Texas Christian University Press.

ANOMALOUS PRIMATES

Collective term for ape-like creatures from around the world that defy explanation. Such creatures as Bigfoot, yeti, almasti, and orang-pendek are cryptids whose ex-

There are a number of ape-like cryptids around the world that fall into the category of anomalous primate.

istence has yet to be proven conclusively through scientific evidence. They are reported form all parts of the world except the poles. Their existence comes primarily from eyewitness reports, apocryphal stories, and disputed physical evidence. Mainstream science tends to reject the existence of these creatures as anything other than folklore, misidentifications, or hoaxes.

See also: Bigfoot; Cryptozoology.

AQUATIC APE THEORY

Alternative human evolution theory first put forward by television documentary writer turned evolutionary science theorist Elaine Morgan in *The Descent of Woman* (1972). Aquatic Ape Theory (AAT) holds that, rather than the usually accepted course of human evolution taking place on the savannahs, early hominids went through a semi-aquatic phase where they lived almost exclusively in the lakes and rivers of central Africa. Evidence of this phase is evident in certain aspects of modern human morphology, including relative hairlessness, bipedalism, and sub-coetaneous fat layers. Morgan claimed that humans share these characteristics with dolphins, whales, and other marine mammals, and that the only way to account for this would be if our ancestors had been aquatic. Morgan also argued for a far more prominent role of female behavior in human evolution.

Elaine Morgan came of age in the rebellious 1960s when women around the world were just beginning to take their rightful and previously denied place in modern society. She saw much of the pain and suffering in the world being the product of selfish, uncaring, and war-like male behavior. In 1967, she read Desmond Morris's popular book on human evolution, *The Naked Ape.* Fascinated by the book, she contacted Morris who put her onto the scholarly work of Oxford University marine biologist Alistair Hardy (1896–1985). Hardy had written several articles on his idea of an aquatic phase in human evolution. Immediately drawn to the idea, Morgan adopted Hardy's work, gave it a feminist spin, and dubbed it the Aquatic Ape Theory. Hardy's idea was not new, however. German scientist Max Westenhöfer had put forward an idea very much like AAT in 1942 in *The Unique Road to Man,* but as Nazi scientists were being shunned, his work did not receive wide notice.

In addition to the marine morphological aspects of human anatomy, Morgan argued that human evolution was largely the product of females and the way they raised young. She rejected the hunter-killer view of male-driven human evolution popular at the time. She also disparaged male evolutionary biologists and paleoanthropologists for being male chauvinists in their theorizing. Their work, she claimed, was blinded by a male centered view of the world. She claimed that male scientists were living vicariously as adventurers and Tarzan-like characters when they did their work. She also argued that the male dominated scientific elite were keeping ideas like hers from being heard. Far from stifling her work, most scientists, both male and female, simply ignored what they saw as a biased and politicized position on human evolution put forward by an amateur with little or no training or knowledge of the field.

Other publications followed. These include *The Aquatic Ape* (1982), *The Scars of Evolution* (1990), and most recently *The Naked Darwinist* (2008). By the late 1990s, Morgan's work was receiving some positive review by the mainstream. It was pointed out that not all of her ideas were completely off the mark. The major problem of her work is that she offers little or no physical evidence to support it. Fossil hominid remains have been found near bodies of water, or fossil bodies of water, but none indicate that these hominids were doing anything other than drinking or bathing in the water. Nothing suggests they lived habitually in an aquatic environment.

Further, Morgan is guilty of some of the same missteps as her male targets. While many male evolutionary scientists did let their egos and Eurocentric worldview color their work, she has done the same, except from a female point of view. A case is point is her 2005 book *Pinker's List*. A specialist in human intelligence and its evolutionary origins, Steven Pinker

of MIT, supported the idea first put forward by Enlightenment philosopher John Locke of a *tabula rasa*. Pinker argued that the human brain was born essentially blank, and that an individual's life experiences imprinted upon the mind his or her worldview and attitudes, and that evolution accounts for other aspects of human behavior. Morgan ridiculed Pinker's work as outdated. She had no choice but to take this position. Her AAT is based upon the notion that the way females raise hominid young has been crucial to human development. If the human mind begins as a blank slate, as Locke and Pinker argued, then bad human behavior could be argued to be the result of bad parenting by females. If human behavior was somehow hardwired into the brain, then all that held back the apocalypse was the desperate attempts by females to raise compassionate and nonviolent offspring. This would keep females on the exalted pedestal Morgan placed them on.

If anything, Morgan's work may have helped some scientists (not all paleoanthropologists are male, and many of Morgan's critics are female scientists) to consider more actively the role of females in human evolution. Like any other aspect of biological transmutation there is not one single cause of anything. Human evolution is the product of millions of years of a complex interweaving of environmental conditions, changes to genetic structure, struggle for survival, and both male and female child rearing. The simplistic view that males went out to hunt while females stayed around the fire to cook and raise children has been discarded by evolutionary science. Morgan's continued critique of "Tarzanists" in evolutionary biology is less a reasoned scientific discussion than a distracting pseudoscientific politicizing of an important issue.

Further Reading

Morgan, Elaine. 1982. *The Aquatic Ape*. New York: Stein & Day.

Morgan, Elaine. 1990. *The Scars of Evolution*. London: Souvenir Press.

Morgan, Elaine. 1995. *The Descent of the Child*. Oxford: Oxford University Press.

Morgan, Elaine 1997. *The Aquatic Ape Hypothesis*. London: Souvenir Press.

Morgan, Elaine 2008. *The Naked Darwinist*. n.p.: Eildon Press.

ARYAN MYTH

Cultural and occult tradition that holds that modern Germanic people are the racial descendants of the Aryans, a supposed "master race" that initially arose in Central Asia in ancient times and then spread to Europe. The Aryans are believed to have been biologically, intellectually, physically, and culturally superior to all other ethnic groups. The stereotypical Aryan is a tall, well-built, blond-haired, blue-eyed individual. Aryan theory became popular in late-19th- and early-20th-century western Europe, particularly in France and especially Germany and Austria (a British variation on the theme focuses on the Anglo-Saxons). It reached its peak in the racial ideals of the Nazis, but has continued to survive to the present within the world of modern-day white supremacy. It is a combination of pseudoscience and pseudoarchaeology, mixed with a healthy dose of anti-Semitism and race hatred.

There are two categories of Aryans, the genuine ancient Indo-Germanic people of prehistoric times, and the mythical Aryans. The former did exist as a people, but by the time of the Greco-Roman period had intermingled so much with the local populations they encountered as they migrated westward, they vanished. The latter only existed in the imaginations of Nazis and white supremacists (the Anglo-Saxons were a distinct people and not a figment of the imagination).

By the latter part of the 19th century, the romanticism of German philosophers like G. W. Hegel was revived into a new romanticism that was metaphysically based. At the same time, German naturalist Ernst Haeckel's conglomeration of nature worship, the occult, and science gave supposedly rational support to the notion of recreating the German man into a new being in tune with nature. Under Haeckel, the Monist League sought to bring an awareness of the importance of biology and the dangers of racial decline brought on by miscegenation (race mixing).

As a result, science and the occult began to merge into a hybrid. Germany's political upheavals and economic crises had intellectuals searching for a unifying principle around which the German people could rally, and the study of human evolution contributed to that unifying history. Many German intellectuals became obsessed with their supposed Aryan origins. This philosophical trend was partly a result of the German middle class fear of slipping into the obscurity of lower-class status brought on by economic hardships. Aryan mysticism, the occult, and ideas about man's central-Asian origins all seemed to prop up their social status, allowing them to argue their superiority. German philosophers and kindred spirits saw themselves as surrounded by degenerates who were trying to destroy them and their culture. The people at the focal point of this worry were the Jews. Scapegoats are necessary for nations in crisis, and the Jews of Europe made the easiest target. Their supposedly inferior evolutionary origins and biology became the hook on which many hung their fears and hatreds.

In this climate, pro-German social philosophies flourished. The Volkish movement, an offshoot of German romanticism that first appeared in the 1870s, glorified the mystical oneness of the German people, or *Volk,* and sought personal fulfillment through exposure to nature. Volkish proponents rejected Christianity as not addressing the specific concerns of Germans and argued for a cultural soul directly connected to nature in a way unlike other races (especially the Jews). Volkish theory emphasized a natural spirituality that thrived on race-based imagination, creativity, and a superior biological origin. Under Volkish thought, Aryans were believed to have been a race of supermen who evolved in central Asia and then spread out to conquer the world, and the modern Germans were their direct descendants. After popular mystic Madame Blavatsky's (1831–1891) *Secret Doctrine* (1888) was translated into German in 1901, it quickly became a blueprint for budding German mystics. Occult evolution is about advancing, not only physically, but spiritually as well. Progressive levels are attained to lift the initiate above the pale of everyday existence and allow the lower orders to reach a superior position in the cosmos. Spiritual evolutionary struggle was thought to make a better spirit in the same way it was thought that physical evolutionary struggle makes a better organism; Aryans, Volkish thinkers believed, were the result of this process.

The preoccupation with mysticism in Germany was epitomized in the life of Guido Von List (1848–1919). A third-rate, would-be aristocratic, Von List created a mythical personal history as well as a history for the Aryan German nation in a se-

ries of flowery, pseudo-Renaissance style novels. The occult world he created was inspired by German mythology, the music of composer Richard Wagner, Madame Blavatsky's evolution theory, and occultism, and it was full of mist-shrouded castles, dark and foreboding forests, magical powers, and was populated by beautiful, heroic, and racially superior Aryan *übermenchen* engaging in epic adventures. (Critics have chortled that J.R.R. Tolkien's *Lord of the Rings* trilogy is a Britannic version, where the bravest and smartest characters are all pasty white.)

In addition to individuals like Von List, groups such as the Germanen Orden and the Thule Society, all of whom ascribed to the Aryan myth, flourished. The hodgepodge of occult spectacle, the bastardization of evolution theory, and anti-Semitism that is Aryan mythology were adopted by many of the key members of the early Nazi Party as the basis for their new religion. Under Hitler, Nazi ideology embraced Aryan biology. Hitler's close ally Heinrich Himmler constructed the SS, Hitler's personal bodyguard, around the pseudohistory of Aryan evolution, as mapped out in the works of Von List and others. Himmler also created a special archaeological unit, the *Ahnenerbe* (Ancestral Heritage Society), expressly to search for "scientific" proof of his beliefs. He sent expeditions to Tibet, Venezuela, and the Antarctic to search for Atlantis and holy relics to aid the Nazi war machine.

Himmler was intrigued by the idea of Aryan Myth and worked it into his philosophy. He made his headquarters for this fringe thinking at the imposing Wewelsburg Castle, turning it into a Nazi monastery. Within its massive, dank walls, Himmler created a fantasy world and dreamed his noxious dreams of Aryan heroes, Holy Grail legends, and world conquest. Taking their lead from American eugenicists, the Nazis added another step—known as the Final Solution—to the process of how to deal with "undesirables" and enemies of the Aryans by putting what would come to be known as the Holocaust into operation.

Besides the fact that there is no physical evidence to support it, one of the ironies of the Aryan myth is that most of its great promoters—Guido Von List, Madame Blavatsky, Heinrich Himmler, and Adolf Hitler—did not live up to the physical ideal of Aryan stature. They were not well-built, blond-haired, or blue-eyed. According to Aryan eugenic ideals, all of them would have been eliminated as unfit. Aryan mythology and its related ideas were little more than a superficial attempt to give scholarly and scientific legitimacy to race hatred and intolerance.

See also: Eugenics.

Further Reading

Goodrick-Clarke, Nicholas. 1993. *The occult roots of Nazism.* New York: NYU Press.
Dhavalikar, MK. 2007. *Aryans, Myth and Archaeology.* New Delhi: Munshiram.

ASTRAL PROJECTION

A technique for creating one's ethereal double and sending it off on travels across space and time. Astral projection is similar to dreams or out-of-body experiences, but where out-of-body experiences happen accidentally as the result of some physical traumatic event, astral projection is done deliberately as part of a religious ritual aiming at exploring other planes of existence. The double, or astral body, is an entity that helps link the physical body to the soul. Believers say the entity can be detached from the body and sent on journeys, or projected, to astral planes, other levels of existence traditionally thought to be located between heaven and earth. Classical writers said these planes were the abode of angels, spirits, and souls of the departed. Religious and metaphysical adepts claim the ability to project their spirit doubles through fasting, meditation, and trance inducement. The idea that a place exists between heaven and earth is common to most religions. The Austrian philosopher and occultist Rudolf Steiner (1861–1925) claimed that, by astral projection, he could read the Akashic Record. A supposed collection of all human knowledge, the Akashic Record is said to exist as either a library or the very mind of God. American prophet Edgar Cayce also claimed to be able to access the Akashic Record through astral projection. Other than anecdotal eyewitness accounts, there is no evidence of the ability to astral project, the existence of other planes, or of the Akashic Record.

Further Reading

Bruce, Robert and Professor C. E. Lindgren. 1999. *Astral dynamics: A new approach to out-of-body experiences.* Charlottesville, VA: Hampton Roads Publishing Company.
Steiner, Rudolf. 1904. *Theosophy: An introduction to the supersensible knowledge of the world and the destination of man.* London: Rudolf Steiner Press.

ASTROLOGY

The study of the movements of the heavens as they relate to and interact with the lives of human beings. Unlike astronomy, the scientific study of the stars, planets, and other celestial phenomena for their own sake regardless of human activity, astrology is a pseudoscience that covers only the patterns formed by the stars and planets as seen from the earth (astrology also does not take into consideration such phenomena as quasars, nebulae, or other similar forms). Astrology tends to treat the heavens in the same fashion as they were treated during ancient times and Middle Ages: as lights attached to the surface of a crystal globe at the center of which hangs the earth. It is believed that the patterns of stars and planets at the moment of one's birth are an indicator of the person's future character and life path. Each month has a corresponding zodiac sign—Cancer, Libra, Taurus, and so on—which also dictates the person's character. "Reading" the arrangement of the stars, a horoscope, or time view allows for divination and prophesy. Astrology may be one of the oldest organized intellectual endeavors of the human experience and is still widely popular today.

Philosopher of science Paul Thagard used astrology as a model for determining the difference between science and pseudoscience. He argued that astrology cannot be tested or falsified, has assertions rather than evidence, is not peer-reviewed for efficacy, does not seem to advance in technique or underlining theory, and has not discarded out-of-date thinking for new ways. While astrologers now employ computer modeling to put together charts and plot the movements of stars and constellations, the underlining theory is still medieval and unchanged.

See also: Alchemy; Divination; Falsification.

Further Reading

Kassel, Lauren. 2007. *Medicine and magic in Elizabethan London: Simon Forman: Astrologer, alchemist, and physician.* Oxford: Oxford University Press.

Thagard, Paul. 1978. Why astrology is pseudoscience, *Proceedings of the philosophy of science association,* (1): 223–34.

Woolfolk, J. M. 2008. *The only astrology book you'll ever need.* Lanham, MD: Taylor Trade Publishing.

ATLANTIS

Legendary island nation said to have been home to an advanced society that existed in ancient times, but was destroyed in an unknown catastrophe. Few mythical places have drawn as much pseudoscientific attention as Atlantis. It has become the archetype of the Golden Age society for speculative writers since it first appeared in written accounts. Thomas More used the legend as a model for his *Utopia* (1516), from which the modern idea and name for a perfect world comes. While entire libraries of books have been written on Atlantis speculating on its location, its culture and meaning, as well as works claiming to have found it, the original source material all of this is based upon is relatively scanty. Countless attempts to find Atlantis have been put forward in various places around the world, but no confirmed location has been determined. There is no confirmation that Atlantis exists outside the imagination.

The point of origin for the huge modern Atlantis literature is in *Timaeus* and *Critias* (both of around 360 BC) by the Greek author Plato. Plato's teacher, Socrates, and his musings on the nature of the perfect state inspired these works. In these writings, presented as a dialogue, Critias says he knows a good example of a perfect state and so tells the story of Atlantis, which is located beyond the "Pillars of Hercules." As such, Atlantis is an imaginary place created by Plato to be used as a model for making a point. Plato may have gotten the original idea from the work of an earlier author, Solon (638–558 BC), who in turn may have brought it from Egypt years before.

Plato's writings fired the imagination of many authors right from the start. Thomas More's Utopia was an island nation in the Atlantic Ocean. Francis Bacon used the myth for his *New Atlantis* (1627), a place where reason, science, and enlightenment would reign. There have been both those who argued Atlantis was a real place and those who argue it is fiction. Isaac Newton speculated on Atlantis in *The Chronology of Ancient Kingdoms* (1728). Later scholars thought Atlantis was either off the coast of the Americas or in the Americas—as Bacon did—possibly linked to Mesoamerican cultures like the Aztecs. Late-19th-century occultists were especially transfixed by Atlantis myths and tied them into a number of other "lost civilization" stories, like Mu and Lemuria. All these ideas and speculations were relatively confined within esoteric circles until late in the century.

The book that helped launch the modern public's fascination with the idea was *Atlantis: The Antediluvian World* (1882) by Ignatius Donnelly (1831–1901). Growing up in Philadelphia, Donnelly's family hit hard times when his doctor father died from typhus in the 1830s. Intelligent and ambitious, Donnelly began studying the law then went into politics. After some hints of financial wrongdoing, Donnelly headed out West. He attempted to form a utopian community in what was then the Dakota Territory, but it failed, leaving him in debt. He went back into politics and eventually became the lieutenant governor of Minnesota in 1860, Republican Party Congressman from 1863–1868, and a state Senator from 1986–1878. He was a supporter of freed blacks' rights as well as women's rights.

After leaving politics, Donnelly returned to his first love, literature and writing. As a utopian, he was fascinated by Plato's description and delved into a period of intense research. The culmination of this period was *Atlantis: the Antediluvian World*. (The next year he released *Ragnarok*, in which he hypothesized about the earth's catastrophic past, which included the devastatingly close pass of a comet.) It became a best seller and was influential to scores of later writers of fiction and nonfiction alike, and remains influential to the present day. He connected later ancient cultures in an attempt to show where the Atlantians escaped to after their own civilization was destroyed by a cataclysm. Many ideas Donnelly worked into the myth system were not mentioned by Plato, yet were later taken as givens by other authors; these include notions that Atlantis as a technologically superior society, that the city's people were the origins of all other races and civilizations, and that they had scientific knowledge now lost to the modern world. Donnelly's work anticipated that of Immanuel Velikovsky, Graham Hancock, and other late-20th-century neocatastrophist, lost civilization, and lost Golden Age writers and believers.

The American psychic Edgar Cayce claimed to have seen Atlantis in some of his many visions and said it would reappear in the 20th century. When a strange block-like formation off the coast of Florida, commonly called the Bimini Road, was discovered in 1968, it was seen by some as support for Cayce's abilities and proof of the existence of Atlantis.

Geologists pointed out that the Bimini Road formation is a rare but naturally appearing geologic formation, not the remains of a lost civilization.

A number of scholars and archaeologists have suggested places around the Mediterranean that contain evidence of past cultures that might have been the model for Plato's Atlantis: the island of Santorini, also known as Thera, being a well-known example. There have been far more outrageous and scientifically dubious candidates. Atlantis has been located by enthusiasts in the Atlantic as far north as Ireland and as far south as the South Pole. In good pseudoscience fashion, modern-day Atlantis hunters use the trappings of marine biology, geology, continental drift, deep-sea drill cores, and other high tech approaches to "prove" where it is. If you were to take all the spots claimed by Atlantis hunters and plot them on a map of the world, you would all but cover the globe. One especially ambitious book, by Brazilian physicist Arysio Nunes dos Santos, *Atlantis: the Lost Continent Finally Found* (2005), claimed not only that had he found Atlantis, but also the Garden of Eden and the "true" location of Troy, all together in the South China Sea. Fortean author Colin Wilson suggests that there is a connection between the inhabitants of Atlantis and the Neanderthals, saying that it was, in fact, the Neanderthals who founded Atlantis. Like the Holy Grail, the lost city of Atlantis represents something special in the hearts and minds of those who pursue it. The search for Atlantis, like so many of the prizes sought after in pseudoscience, is the search for the self. It becomes a mirror held up to the searcher. Admittedly, the objects of genuine science are often the same, the researcher looking for a cure for AIDS, a way to fly to the moon, a way to see to the end of the universe, to explain how animals evolve, or the nature of life are searching for something deeper as well. It is not really about Atlantis—it doesn't matter if Atlantis or Bigfoot exists—what matters is the hunt, the motivation to find out. This is something both science and pseudoscience do have in common.

See also: Hidden History.

Further Reading

Harrison, W. 1971. Atlantis undiscovered: Bimini, Bahamas. *Nature* (230): 287–89.

Joseph, Frank. 2004. *The destruction of Atlantis: Compelling evidence of the sudden fall of the legendary civilization.* Rochester, VT: Bear & Company.

Spence, Lewis. 1924. *The Problem of Atlantis.* London: William Rider and Son.

Wilson, Colin. 2006. *Atlantis and the kingdom of the Neanderthals: 100,000 years of lost history* Rochester, VT: Bear & Company.

ATOMIC MUTANTS

A popular topic of science fiction literature and film, atomic mutants are living things, usually insects, animals, or humans, and less often plant life, that have been radically and suddenly altered in their morphology due to exposure to atomic radiation, often a bomb blast. The heyday of such works was the 1950s and 1960s, though the genre has survived to the present. There are two basic types of atomic mutants: already-monstrous creatures that are released from geologic captivity in rock or ice formations by a nuclear explosion, or those where a nuclear explosion or radiation cause normal organisms to mutate into monsters through enhanced size or the sudden appearance of physical abilities they do not normally possess.

The first film version featuring such an organism was *The Beast From 20,000 Fathoms* (1953). Based loosely upon the short story of the same name, by author Ray Bradbury (and later changed by Bradbury to *The Fog Horn*), the beast, called a Rhedosaur, is set free from an icy arctic tomb by a nuclear bomb blast. It heads to New York City, where it rampages about before being dispatched by the military at the Coney Island amusement park. The success of the film inspired arguably the best of the genre, *Them!* (1954), in which lingering radiation from the first nuclear explosion at Los Alamos causes common desert ants to grow into monsters who then invade Los Angeles where they make pests of themselves. *Them!* is a film noir detective film, but with mutants, where even the title leaves you guessing what the central evil is until the ants make their appearance well into the story. *The Beast From 20,000 Fathoms* (along with the non-atomic mutant *King Kong* [1933]) also inspired Japanese filmmakers to produce *Godzilla* (1954).

An entire menagerie of animals has been mutated by radiation in film. Insects are especially popular. Along with the ants of *Them!*, there are the grasshoppers of *The Beginning of the End* and the praying mantis of *The Giant Mantis* (both of 1957), but there have also been crabs, turtles, and humans. The problem with atomic mutants is that they do not conform to the known laws of science. In the case of bugs, they have exoskeletons. This means that, if they were grown to monstrous size, the square cube law would take effect. As they get bigger, the bugs' ability to move and breathe would be greatly reduced and ultimately prove fatal. The largest animals known ever to have existed, like the brachiosaurs, compensated for their huge size by moving very slowly and being herbivores. Likewise, the blue whale, the largest extant species, can grow to enormous size because the buoyancy of the ocean supports it.

That radiation does have an effect upon the genetic structure of living things was determined in the early 20th century. Natural radiation emitted by the sun and other stars can alter DNA. It is these natural mutations that build up slowly over time and contribute to the process of evolution. Only *Them!* made a passing reference to the ants having grown larger over successive generations of prolonged radiation. Sudden bursts of radiation, either natural or man-made as a result of atom bomb explosions or leakage from nuclear power plants, do great damage to living tissue. Such effects are quickly fatal and would not cause an organism to mutate in the science fiction sense of rapid and profound change. Therefore, getting bitten by a radioactive spider would have killed Peter Parker, not turned him into Spider-Man.

Further Reading

Newman, Kim. 2000. *Apocalypse movies: End of the world cinema*. New York: St. Martins Griffin.

B–E

BACON, SIR FRANCIS

British philosopher often credited with helping bring on the scientific revolution and, in turn, the modern world by articulating a technique for studying the natural world called the scientific method. Born into a family with close ties the monarchy, Bacon (1561–1626) entered Trinity College, Cambridge at the age of 12. He received a rigorous education that entailed the reading of Greek authors, especially Aristotle. This caused him some intellectual confusion, as he greatly admired Aristotle, but not the Aristotelian philosophy of his contemporaries. Medieval scholasticism focused on studying an interpreted form of Aristotle and others and came to lean heavily on the virtual infallibility of those authors. Bacon found this troublesome, as Aristotle himself taught that it was wrong to blindly accept past authorities, and encouraged his students to see the world with one's own eyes and intellect.

After completing his studies at Cambridge, Bacon traveled to France and served as a diplomat and courier while continuing his studies there. Family financial difficulties drove him home to become a lawyer instead of philosopher, as he had hoped. While trying to carve out a career for himself, Bacon began to write about religious and political topics. In 1613, family connections led to his appointment as attorney general, and in 1618 he was named Lord Chancellor to King James I. A corruption scandal found him out of work shortly after, and even imprisoned briefly in the Tower of London.

While his place in British political history is interesting, Bacon is best remembered for his philosophical writings. Taken as a body of work, Bacon's writings put forward a method of how to study the natural world. A religious man, he believed that, far from compromising one's faith, technical, philosophical, and scientific knowledge enhanced one's belief in God. It was in *Meditationes Sacrea* (1597) that he made his well-known statement, "Knowledge is power."

Bacon's most famous work was *The New Atlantis* (1627). A novel rather than a scientific treatise, it was meant to be the first in a line of books in which he laid out his ideas of a perfect society, or utopia. The utopian ideal was based upon rational knowledge gained according to rigorous methods of scientific enquiry. In later works, like *Magna Instauratio* and the *New Organon,* Bacon laid out the process of the scientific method: ideas must be tested through observation and experimentation, then, once confirmed, a further level of inquiry is performed, then another and another until a mountain of trustworthy facts are accumulated. These facts form generalizations about the natural world.

While Bacon was not as specific with the details of his method as later readers would have liked, his core idea of using rigorous standards and procedures of inquiry in which the recording of facts and the constant testing of hypotheses is now the underlining philosophy of modern science and indeed helps define what science is. The idea that the only kind of valid knowledge is that which is the result of the accumulation of data was not completely adopted. Scientists today acknowledge that simple observation alone is not enough; intuitive leaps are required, based upon facts, to formulate theories and hypotheses as well. Later philosophers added important ideas about falsification, paradigms, and other methods that Bacon did not discuss. The result is that, while Bacon is held up as a pivotal figure in the scientific revolution, little modern science is based solely on pure Baconian logic, rather it is a far more complex exercise. Bacon is held alternately as a shrewd thinker who helped create the modern world and as an insufferable self-promoter who produced no real science himself, but was, as the German philosopher G. W. Hegel called him, a "coiner of mottoes."

Further Reading

Peltonen, Markku. 2008. *The Cambridge companion to Bacon.* Cambridge: Cambridge University Press.

BERMUDA TRIANGLE

Region of ocean located in the Atlantic and anchored in the Caribbean Sea in which ocean-going vessels and aircraft disappear in large numbers under mysterious circumstances. Articles about the region have appeared since at least the 1950s in popular men's magazines like *Fate.* By the 1960s, the area was being referred to as the "Deadly Triangle." An important early article on the disappearance of ships, and especially planes, in the region was in 1964 in *Argosy.* Titled "The Deadly Bermuda Triangle," by author Vincent Gaddis (1913–1997), the article discussed the contentious story of Flight 19 and coined the term Bermuda Triangle. The 1970s saw a series of publications on the Bermuda Triangle, also now referred to more provocatively as the "Devil's Triangle," in articles and books. One of the most popular was *The Bermuda Triangle: An Incredible Saga of Unexplained Disappearances* (1974) by esoteric subjects author Charles Berlitz (1914–2003). A skilled linguist whose family published a long line of popular language instruction books, Berlitz was interested in unusual subjects and began researching losses of ships and aircraft. Drawn to the Bermuda Triangle mystery, he noticed a pattern of hard to explain incidents and legends about doings off the coast of Florida. He saw that the region had a history of unsolved disappearances and high profile cases. He listed many of them and focused on ones that could not be explained by natural means. This resulted in the conclusion that something mysterious was going on in the region.

Researchers scoured record books, ship's logs, police and Coast Guard reports, and local legends and found many more "unexplained" cases. Because some researchers could not explain some of these cases naturally, they resorted to supernatural explanations. The large number of lost ships and aircraft in the triangle—which has been alternately mapped with most of its southern region in the Caribbean and off the coast of Florida and its upper most portion being placed as far north as New Jersey or even Ireland—led some to put forward ideas of ghost pirates, magnetic anomalies, "electric fog," astral vortices, sunken cities, and, of course, UFOs and USOs (Unidentified Submerged Objects). Atlantis is reputed to have been in this region, as evidenced by the existence of the Bimini Road.

The most famous case of the Bermuda Triangle is that of Flight 19. A key publication on the case was by magazine writer Allan Eckert and appeared in *American Legion* magazine in 1964. On December 5, 1945, a group of five U.S. Navy Avenger torpedo bombers out of a Florida airbase were on a mission in the skies over the Atlantic Ocean. Having completed their training, they were heading home when the flight's leader Lt. Charles Taylor became disoriented. Air controllers in Fort Lauderdale tried to talk the flight back home, but they had become hopelessly lost. In radio communications with controllers, Taylor seemed upset, confused, and increasingly panic-stricken. Eventually radio contact was lost and the flight disappeared. A massive search by ships and aircraft failed to find a single trace of Flight 19. Adding to the mystery, a large flying boat that was part of the rescue mission also disappeared without a trace. In his article, "Mystery of the Lost Patrol,"

Eckert reported Lt. Taylor saying that "nothing seems right," and other enigmatic phrases. Eckert linked the case to supernatural phenomena and made a number of claims about the incident that were found to be untrue. He said the flight took place in good weather when in fact the flight took place in terrible weather, with dark and stormy skies and rough seas. He also claimed Taylor was a skilled navigator. Though he was a combat veteran, Taylor was not known as a skilled navigator. He had, in fact, ditched in the Pacific twice during World War II, reportedly as a result of his a poor navigation skills.

Author Lawrence David Kusche contradicted all the claims of the unearthly made about the Bermuda Triangle. In *The Bermuda Triangle Mystery-Solved* (1975), and again in *The Disappearance of Flight 19* (1980), Kusche took issue with how these cases were presented. He took special umbrage with Charles Berlitz, accusing him of incompetence and even fabricating evidence. Kusche carefully examined the naval records, interviewed eyewitnesses, and, as a pilot himself, flew the route of Flight 19. He argued that while tragic, the loss of the planes could be put down to mechanical and pilot error. He also asserted that none of the "mysteries" of the Bermuda Triangle could be supported with evidence and were only hearsay, innuendo, and the work of fabulists. He concluded that the Bermuda Triangle was a manufactured mystery with no basis at all in fact.

Others tended to agree. The famous shipping insurance agency Lloyd's of London—whose operations can be traced back to the 1680s—does not charge higher rates for ships traveling through the Triangle. The United States Coast Guard also does not see anything unusual about the number of vessels lost in the Triangle. Some researchers and marine insurance specialists point out that the numbers of losses of either ships or aircraft in the region commonly thought to encompass the Bermuda Triangle are no higher or mysterious than at any other spot in the world's oceans.

The Triangle still has its supporters. Anomalous phenomena investigator Gian Quasar took issue with Lawrence Kusche's debunking. He claims that Kusche makes arguments against the Triangle that he does not support, that he gets his information wrong, uses faulty sources, and generally does just as bad a job in his writing as he claimed about Charles Berlitz. Quasar is skeptical of skeptics and charges that they recycle false reports and supposed facts against the subject and do a poor job of researching the topics they wish to undermine. This helps to illustrate what can happen when one supports a side in a discussion of a pseudoscientific topic. Whether supporting or debunking something anomalous, one has to base his opinions on facts known to be accurate.

See also: Debunkers.

Further Reading

Gaddis, Vincent. 1965. *Invisible horizons: True mysteries of the sea.* New York: Ace.

MacGregor, Rob. 2005. *The fog: A never before published theory of the Bermuda Triangle phenomenon.* Woodbury, MN: Llewellyn Publications.

Quasar, Gian. 2005. *Into the Bermuda Triangle.* Thomaston, ME: International Marine/ Ragged Mountain Press.

BIGFOOT

Also called "sasquatch," Bigfoot is an anomalous primate said to inhabit the Pacific Northwest of the United States and Canada. Reported to be roughly eight feet tall and weighing 500 pounds or more, Bigfoot is a hairy, man-like, bipedal creature of mostly nocturnal and solitary habits. The creature is notoriously hard to find and has appeared in a small number photographs, all of which are contentious as to their authenticity. The most commonly found evidence for Bigfoot and similar forms are the large footprints they sometimes leave behind. Researchers have collected an enormous number of plaster casts of such prints and use them to argue for the creature's existence. The most well known image of Bigfoot comes from the 16 mm film footage commonly called the Patterson-Gimlin Film, which was shot in October 1967 in the Bluff Creek region of northern California by a pair of cowboy/ranchers, Roger Patterson and Robert Gimlin, as they explored looking for the creature.

There are reports from Native American legends of Bigfoot-like creatures that preceded the arrival of European settlers. There are also scattered reports from explorers and trappers in the 18th and 19th centuries who encountered creatures that fit the Bigfoot mold. These reports, and the idea of Bigfoot, remained little known outside of folklorists and local inhabitants until the middle of the 20th century. In 1958, a construction gang pushing a road through the Six Rivers National Forest in northern California, near a stream called Bluff Creek, encountered unusually large footprints in the dirt near their heavy equipment. While they never saw who or what left the tracks, the workers claimed the tracks reappeared often. One of the gang, Jerry Crew, made a plaster cast of one of the tracks and brought it to local *Humboldt Times* reporter Andrew Genzoli. Crew told the reporter about what had happened and that the construction workers had taken to calling the mysterious visitor "Bigfoot." Genzoli put the story on the AP wire service, a move that resulted in an instant sensation. The name Bigfoot carried a great pop culture punch. A number of amateur naturalists already interested in man-like monster sightings began to investigate the Bigfoot incident. Men like Canadian John Green, Swiss-Canadian Rene Dahinden, Irishman Peter Byrne, and many others researched the phenomena, talked to eyewitnesses, and collected evidence about these creatures. With the Bigfoot story, the floodgates opened and many people in the United States and Canada began to come forward telling how they, too, had seen the creatures. Along with the amateur naturalists, a small coterie of professional scientists also joined the search for these elusive animals. These included British primatologist John Napier (1917–1987), whose *Bigfoot: The Yeti and Sasquatch in Myth and Reality* (1973) was the fist major scholarly work to address the evidence, and American paleoanthropologist Grover Krantz (1932–2002), who became the scientist most associated with the study of anomalous primates.

Krantz, from Washington State University, spent most of his 30-plus year career working to articulate an acceptable theoretical paradigm that would link Bigfoot to the extinct Asian primate *Gigantopithecus.* Known from fossil teeth and jaw parts, *Gigantopithecus* is the largest primate ever known. The available fossil evidence suggests to some, though not all, scientists that *Gigantopithecus* was a large biped that weighed in the 500-pound range and stood about eight feet tall. This description is virtually indistinguishable from reports of sightings of Bigfoot. Krantz put together technical data on Bigfoot from the sightings and from the Patterson-Gimlin Film to produce a biomechanical

The skull of Gigantopithecus. *This fossil primate from Asia is thought to be the progenitor of the Yeti and Sasquatch.*

description of the creature that made sense anatomically. He attempted to use the techniques and methodologies of physical anthropology to explain what the creature looked like and, more importantly, to explain why it could exist. One of the key pieces of evidence Krantz relied on were the "Cripplefoot" tracks found near the towns of Bossburg and Colville in Washington state in November 1969. These prints intrigued Krantz because of anatomical anomalies in the right foot that he felt could not have been created by hoaxers. This approach became a common one amongst enthusiasts: if the anatomical details of tracks could not be faked, some of the tracks must be genuine, thus the creatures must exist. In more recent years, anthropologist Jeffrey Meldrum of Idaho State University has taken up Krantz's mantle as the most visible professional scientist to argue for Bigfoot's existence in North America.

Many in mainstream biology and zoology feel that the evidence for such

The Scottish naturalist Ivan Sanderson is one of the cofounders of cryptozoology.

Bernard Heuvelmans (1916–2001) and Ivan Sanderson (1911–1973) went to see the exhibit and became convinced of its authenticity. They both published papers on the find which drew the attention of the Smithsonian Institution, in the form of John Napier, and later, the FBI. The law enforcement agency became interested in the Iceman when it was thought that the creature had been dispatched with a firearm. The FBI's interest was short-lived, but Hanson became worried and took the exhibit down for some time. When it reappeared, Hanson claimed he had been exhibiting a fake all along. The creature and Hanson soon slipped into obscurity.

The most recent Bigfoot hoax was the "Bigfoot in a box" sham, put forward by a pair of Georgia men in August 2008. They put up photos of a carcass they claimed to have found while out hunting. They said they had scooped up the remains and put them into a large box. Much anticipation was generated, and the discoverers claimed they had sent some of the remains for DNA analysis, whose results they would announce at a major press conference. Some eagle-eyed Web surfers saw from the photos that the "carcass" was actually an off-the-shelf gorilla costume with some possum entrails strewn over it. The fraud quickly fell apart and the promised press conference came to nothing.

As with cryptozoology in general, anomalous primate studies—like the Loch Ness Monster, Chupacabra and others—can be settled with the discovery of actual remains that stand up to scientific scrutiny.

Despite the false starts, hoaxes, and general lack of enthusiasm from professional scientists, Bigfoot researchers continue to go into the woods in hopes of finding evidence of their quarry. Since the 1950s, sightings have continued and have come

creatures is slim and circumstantial at best. They argue that such an animal runs counter to evolutionary theory and that the attempt to link Bigfoot to *Gigantopithecus* is dubious, as the most recent fossils are 300,000 years old. In addition, no *Gigantopithecus* fossils have ever been found in the Americas. Another problem with the field is the proclivity for hoaxing that goes on within it. Following his death, Ray Wallace (1918–2001), the foreman of Jerry Crew's construction gang, was accused of having faked the famous Bigfoot tracks. His family stated that he had been faking tracks most of his adult life. Other researchers who were familiar with Wallace accepted that he was a hoaxer but argued he had not faked the tracks Jerry Crew found.

The field of anomalous primate studies is studded with famous frauds, including the Minnesota Iceman of 1968. Carnival exhibitor Frank Hanson was showing a body of a hairy man-like creature in a block of ice that he claimed was a "missing link" he had acquired from an anonymous millionaire. Pioneering cryptozoologists from all over the country, from Florida to New York, to Oklahoma and Texas. Now armed with high tech gear like night vision scopes, heat sensors, and motion triggered cameras, the Bigfoot hunters still head out into the woods like bird watchers. If creatures like Bigfoot are ever found it will not be by professional scientists, but by the amateur enthusiasts.

See also: Anomalous Primates; Cryptozoology.

Further Reading

Coleman, Loren. 2003. *Bigfoot: The true story of apes in America.* New York: Paraview Pocket Books.
Daegling, Dave. 2005. *Bigfoot exposed: An anthropologist examines America's enduring legend.* Lanham, MD: AltaMira Press.
Green, John. 2006. *Sasquatch: Apes among us.* Blaine, WA: Hancock House.

BLIND TESTING

Science, as a materialistic enterprise for learning the nature of the universe, needs ways of checking its facts and experimental results for accuracy. Experiments must be reproducible by others and results only arrived at once or which are not testable are suspect. The most respected tests are known as "blind." The term blind comes from the way the tests are set up so that experimenters and test subjects do not know what they are testing and therefore focus on performing the test in an unbiased way. Scientific research is expected to be objective, however scientists are human and no matter how hard they try to be objective, their personal characters can sometimes alter a test; an experimenter or test subject can consciously or unconsciously favor one result or another. By making the test blind, the "experimenter factor" is eliminated. The most common way of do-

ing blind testing is to conceal the identity of test materials; this also includes the use of a "control" subject that has not been specially treated. Using a control gives the test a set result with which to compare other outcomes. Other ways to achieve a blind test are to ensure that all testing equipment is uniformly calibrated and to verify that any defects in the physical test equipment do not alter the results. Experimental data that passes blind, double-blind, and even triple-blind tests are considered genuine and highly respected. Pseudoscience subjects are rarely subjected to blind studies, and when they are they regularly fail.

CARDIFF GIANT

Supposed giant man found in New York, which caused some interest until it was discovered to be a hoax. In November 1869, a farmer named Newell dug up a large, bizarre fossil on his land outside Cardiff, New York. The story quickly spread through the media to the outside world. The fossil was a large figure of a man almost nine feet tall who had turned to stone through some unknown but extraordinary process. The farmer put up a tent and began to charge admission to those who came to look at it, though he would not let observers get too close. Protestant ministers seized on the fossil as scientific support for the accuracy of scripture. One New York geologist, James Hall, came to look and the statements he made about it being intriguing were taken out of context by newspapers as scientific support for the giant's genuineness. Eventually, a young Yale University paleontologist, Othniel Charles Marsh (1831–1899), who would go on to a long and distinguished career, inspected the anomaly and quickly saw that it was a block of gypsum carved to look like a human body. Marsh's assessment that the Cardiff Giant, as it was known, was a hoax did not deter showman P. T. Barnum who, unable to buy the humbug, simply made one himself and put it on tour. A few years later, a similar giant was found in County Antrim, Northern Ireland. Found by a Mr. Dye, the Irish Giant looked like the Cardiff Giant and was exhibited in Dublin and other Irish cities. It, too, eventually drifted into obscurity.

Further Reading

Tribble, Scott. 2008. *A colossal hoax: The giant from Cardiff that fooled America.* Lanham, MD: Rowman & Littlefield.

CHUPACABRA

Cryptid reported to inhabit the island of Puerto Rico. *El chupacabra* has been reported in the United States, but sighted mostly in Mexico and other Latin American countries. First reported in Puerto Rico in 1995, the name means "goat sucker," and derives from the creature's apparent taste for farm animal blood and its vampiric way of obtaining it. The few eyewitnesses who claim to have seen the creature describe a strange composite of dog and reptile with spiky projections along its back, glowing eyes, and claws. A number of recent incidents where the creature has been reported captured or filmed have, upon close inspection, shown

A Chupacabra in a hurry. This Latin American cryptid was first reported in the 1990s.

only sickly wild dogs, coyotes, or foxes. Skeptics argue that all the behavior reported associated with the *chupacabra* can be explained by the behavior of wild dogs known to inhabit the areas of the sightings. This creature may be more local cultural product rather than genuine cryptid; however, recent sightings have been reported from North America and Russia.

See also: Cryptozoology.

Further Reading

Coleman, Loren and Jerome Clark. 1999. *Cryptozoology A to Z: The encyclopedia of Loch Monsters, Sasquatch, chupacabras, and other authentic mysteries of nature.* New York: Fireside.

COLD SPOTS

A phenomenon reported by ghost hunters and paranormalists, cold spots are areas of sudden and unexpected temperature decrease, and often held to be indications of recent activity by spirit entities (i.e., ghosts). Spirit investigators employ temperature gauges to search for and isolate these cold spots. The temperature gauge is used in conjunction with an electromagnetic field (EMF) meter, ion counter, and other high-tech gear. EMF meters are used to detect electromagnetism, while ion counters are used to indicate a rise in ion discharges. These pieces of equipment are considered essential apparatus in any self-respecting ghost hunter's toolbox.

There are a number of operational problems with cold spots. First, it has been pointed out that any location or environment, whether inside or outside, will have natural temperature fluctuations. The air movement that brings on these fluctuations is often undetectable by unaided hu-

man senses; even rapid temperature variations can be caused by the flow of air through a building that, to the unaided senses, has no flowing air. Paranormal investigators have yet to devise a technique for distinguishing between normal airflow and temperature fluctuations and that caused by spirit activity. Second, there is no known and scientifically verified correlation between spirit activity and air temperature. This is, in part, because there is no scientifically confirmed evidence or definition of what a spirit entity is, or if they exist.

See also: Ghost Hunters; Ghost Hunting.

Further Reading

Nickell, Joe. 2006. Ghost hunters. *Skeptical Inquirer* 30 (5).

CREATION SCIENCE

The religious belief that the traditional creation story of the Bible can be supported as literally true by scientifically gathered evidence from disciplines such as biology, geology, physics, and astronomy. Early works that could be called creation science appeared in Britain in the 18th century, but the modern more vigorous variant appeared in the United States in the late 19th century. Creation science practitioners argue that the scientific evidence that mainstream scientists use to support evolutionary thinking in fact supports a creationist point of view. While there are a number of different schools of creationist thought, most creation scientists believe the earth is no more than 10,000 years old, Intelligent Design proponents being an exception.

The modern, Western scientific tradition began during the Enlightenment, but well into the 19th century there was little conflict between religion and science. Most practitioners of the fledgling biological, and especially geologic, sciences were interested in reconciling the physical evidence of the earth with scripture. There are many and complex reasons for the separation to have begun, but a major step toward the parting of science and religion was the introduction of the theory of evolution in the mid-19th century. It was with the separation of science and religion that creation science was born.

An attempt to bridge the gap between evolution and religion can be seen in the work of George Frederick Wright (1838–1921). He studied for the ministry at Oberlin College in Ohio during the time when the college's president was the legendary Charles G. Finney (1792–1875), a prime mover in the religious movement known as the Second Great Awakening. Finney stressed the idea that the Bible was the ultimate authority and that religious belief was supported by evidence from nature. Becoming a minister upon his graduation, Wright began to teach himself natural history by reading the works of Charles Darwin and others, whose books' central themes were in direct contradiction to what he had been taught as a seminarian. By the 1880s, however, Wright grew more conservative after taking a position at his old school, Oberlin, and changed from defending science from religion to defending religion from science. He grew to fear that science, and evolutionary thinking in particular, were seeking not just to coexist with, but to undermine, religious faith.

In time, Wright lost his liberal view and adopted an increasingly literalist stance. He wrote a number of books on natural history which received poor reviews from scientists. They accused him of badly misreading the geology he discussed and dismissed him as a theologian out of place trying to explain science. The stinging critique by geologists pushed Wright deeper into the traditional anti-evolutionist camp. Wright then tried to find scientific explanations for events like Noah's flood and the destruction of Sodom and Gomorrah. Wright's attempt to support religion with scientific thinking and geological and archaeological evidence was a step toward what would eventually be known as Creation Science.

In the early part of the 20th century, a group of loosely affiliated preachers and Christian intellectuals in the United States and Great Britain (where similar concerns had been percolating) joined together and wrote what became the manifesto of a new religious movement. Taking their name from the book's title, Christian fundamentalists began to exert pressure where they could to re-create American society. One of the things they reserved their wrath for was evolution and science in general. Following the Scopes Monkey Trial of 1925, some anti-evolutionists stepped away from a purely scriptural point of view and began building what they thought would be a scientific refutation of evolution. Creation science was the fundamentalist belief in a literal interpretation of the Genesis story of a six-day creation week occurring no more than 10,000 years ago, which all could be proven by scientific methods. These thinkers were known as Young Earth Creationists (YEC), in reference to the calculations of Irish cleric Archbishop James Ussher (1581–1656), who placed the year of creation at 4004 BC. Not all creationists took the Young Earth position. Some allowed the earth might be billions of years old. Gap Theory allowed that, after the creation week, a period of indeterminate length passed before the historical period began. Strict literalists rejected this notion because it did not fit biblical accounts and contradicted God's word. Gap Theory traces back at least to Scottish theologian Thomas Chalmers (1780–1847) who used it as a way of reconciling scripture with the knowledge geologists were producing about the nature of the earth. Along with Gap Theory was the Day-Age Theory that argued that the days of Genesis were not 24-hour days, but were of indeterminate length. Again, strict creationists saw this as a cop-out to try to get around scientific evidence.

In the 1930s, the first in a series of societies was formed to keep the work going. Groups like the Religion and Science Association and the Deluge Geology Society came into being and, though small, were hotbeds of creationist activity and debate. Squabbling among believers grew intense, partly as a result of younger enthusiasts obtaining conventional science educations. With their new knowledge and what they saw in the field, they found it increasingly difficult to reconcile some of the more extreme claims of older followers. The new generations were not quite armchair theorizers like their unlettered forebears; their deeper understanding of the natural world gave them a different insight. Unable to overcome the validity of biology and geology, some creationists began giving ground, accepting scientific tenets as long as they could hold on to the ultimately divine origin of humans and their souls.

The man often credited with having provided the framework for the revival of the creation science movement of the 1960s is Henry Morris (1918–2006) and his book *The Genesis Flood* (1961). A trained engineer with a doctorate in hydrodynamics, Morris, like many of the more scientifically trained creationists, came to

fundamentalism later in life. Following work in a series of teaching positions, Morris decided to turn his technical training to the problem of explaining Noah's flood. He came to believe that traditional geology and evolution were the enemies of religion. His literalist interpretation of the Bible became increasingly inflexible and he refused to concede to anything that did not directly support scripture. To this end, he joined forces with theologian John Whitcomb to produce *The Genesis Flood*. Abandoning uniformitarian geology for catastrophism, Morris argued that an envelope of water vapor surrounded the earth in Biblical times and that God piercing it was what provided the waters for the deluge. After going to great lengths to employ the language of fluid dynamics to explain the flood, he reached an impasse that he surmounted through the use of miracles, but in so doing, Morris stepped into the same hole many literalist creation science proponents did. Wanting to beat the heathen scientists at their own game, Morris could only go so far down that road before being forced by simple logic to abandon science and untie the Gordian Knot with a miracle, thus undermining their effort.

Morris's thesis was not particularly new. He was, to a great extent, repeating the work of George McCready Price from the early part of the 20th century. Price (1870–1963) was likely the most influential scientific creationist of the early 20th century and helped pave the way for later writers like Morris. Price grew up a millenarian in a Seventh-day Adventist church in New Brunswick, Canada. One leader of the Adventist church was the charismatic Ellen White (1827–1915), who claimed that God passed information to her followers through her while she was in a trance. She stressed a literal belief in

Genesis without any of the nuanced interpretations like the Day-Age and Gap theories.

In the 1890s, Price briefly attended a small New Brunswick school, taking a few classes in natural history (these were the closest he would ever get to formal training in anything approaching science). By the turn of the century, he began to fixate on the geologic aspects of evolution as the key to the whole system. He paid particular attention to stratigraphy and the explanation for why younger layers were occasionally found below older ones. Geologists had long argued that the strata of the earth went in a progression from youngest at the top to oldest at the bottom. The concept of uplift was used to explain why the sequence was not always so simple: geologic forces could sometimes lift older layers above younger ones, making them seem out of sequence. Price found this explanation unacceptable. As to the initial formation of the layers, Price found Charles Lyell's steady-state, uniform-action explanation also unacceptable. His Adventist upbringing led him to Georges Cuvier's catastrophism. Price modified Cuvier's many periods of catastrophic upheaval and reduced them to Noah's flood, as Morris would do decades later. It was the flood, Price believed, that accounted for the geology of the earth and the fossils that were remnants of those animals that did not get into the ark. He called this system "flood geology" and published it as *The New Geology* (1923).

Henry Morris had the good fortune to address these same issues at the moment when the resurgent creationist movement was looking for a secular explanation to support their position. Morris rode the wave of his success by forming first the Creation Science Research Center (CSRC) and then the Institute for Creation Re-

search (ICR), both of which became influential centers of Young Earth creationism and the heart of the creation science movement. They took on the task not so much of convincing unbelievers in the mainstream scientific community but in influencing sympathetic members of local schoolboards to begin working creation science into their curricula. This was a risky move, as it forced a number of court cases like *Epperson v. Arkansas* (1965) and *Edwards v. Aguillard* (1987), which outlawed the teaching of creation science in the public school system.

Another father of modern creation science is Duane Gish who toured, lectured, and wrote extensively on the cause of creation science and helped to bring it back to life. He earned a doctorate in biochemistry from the University of California at Berkeley in 1953, and worked as a corporate chemist and researcher until 1971, when he joined Christian Heritage College, an institution founded the year before. The next year, the college changed its name to the Institute for Creation Research. Gish eventually became the institute's associate director and vice president. His favorite target is the fossil record. *Evolution: The Challenge of the Fossil Record* (1985) is representative of Gish's work and one of the most quoted works in the genre. Originally published as *Evolution: The Fossils Say No!*, it was an examination and undermining of the idea that fossils support the evolution model. He followed a common creationist technique of taking quotes by evolutionists out of context to argue that what they said did not make sense and that they themselves did not deem evolution to be proven. For example, when influential evolutionist George Gaylord Simpson said it was difficult to find fossils of the Precambrian era even though the sediment suggested they should be there, Gish argued that this meant the evolutionists were condemning their own idea. Differences of opinion or unanswered questions became "proof" that evolution was a false doctrine. Gish used many of the standard arguments of creation science thinking: there were no transitional forms in the fossil record; evolution was not really science but was just as faith-based as any other religion, and fossils actually proved the creation, not evolution, model. Gish also argued that evolutionists believe and still teach the "molecule-to-man" theory. This approach, which argued evolution was a goal-oriented process leading to the advanced form of man, was popular in the 19th century but has long since been abandoned by mainstream paleo-anthropology. Gish lamented that "it is incredible that the molecules-to-man evolution theory be taught as a fact to the exclusion of all other postulates."

Creation science tries hard to cloak itself in the trappings of science, but it is one of the most clearly pseudoscientific concepts in this book. It is based on the premise that the universe was divinely created, but then tries to shoehorn in genuine scientific material to support that premise. Like many anti-evolution concepts, proponents of creation science spend most of their time arguing against science and evolution instead of arguing for what they believe. None of its tenets holds up to testing under the scientific method. Ultimately, it searches for evidence of God, and that alone is enough to take it out of the realm of genuine science and place it firmly in the boundaries of metaphysics.

See also: Anomalous Fossils; Intelligent Design.

Further Reading

Conkin, Paul. 1998. *When all the gods trembled.* Lanham, MD: Roman and Littlefield.

Morris, Henry. 2000. *The long war against God.* Green Forest, AR: Master Books.

Numbers, Ronald. 1992. *The creationists.* Berkeley: University of California Press.

Regal, Brian. 2002. *Henry Fairfield Osborn: Race and the search for the origins of man.* London: Ashgate Press.

Regal, Brian. 2004. *Human evolution: A guide to the debates.* Santa Barbara, CA: ABC-CLIO.

Scott, Eugenie. 2004. *Evolution vs. creationism: An introduction.* Westport, CT: Greenwood.

CROP CIRCLES

Large patterns made in fields of growing crops under mysterious circumstances. The patterns usually appear in wheat fields during the period of greatest growth prior to harvesting, and range from simple circles to highly complex patterns that mimic computer-generated designs. Some anomalous phenomena enthusiasts argue that the circles are made by an extraterrestrial agency and that they portend some great coming event or attempt by that agency to make itself known to mankind. Others argue for various vaguely referenced natural earth energies. They have also been known to appear in snow and ice formations, though only rarely. First noticed in the early 1970s, they have been seen around the world, but are concentrated in the British Isles.

Colin Andrews coined the term "crop circle" in the late 1970s. They became part of popular culture by the 1980s and hit their peak of public interest in the 1990s. When first noticed, they were simple circular formations formed around a central point. The patterns are formed by crunching down the stalks all in the same direction so that they are flattened, but not chopped, to the ground. They eventually evolved into more complex arrangements with multiple circles sometimes connected by lines, until they began to take on extraordinary patters of great imagination, complexity, and beauty. One of the more mysterious elements of the phenomenon is that, despite them appearing in public places, no one is ever seen making them; they seem to appear out of nowhere.

There are scattered accounts of crop circle–like formations back to the 17th century. A single pamphlet with an illustration of a "mowing devil" using a scythe to cut down wheat stalks in an ovoid pattern is often put forward by researchers as proof of the ancient nature of the phenomenon. Some UFO reports include mentions of circular marks being left in grass and dirt near their supposed sightings. Cerealogists (crop circle investigators) report unusual damage done to the crop stalks, including mass bending at the same point without breaking the stalk and exploded "nodes" of the stalk that were caused by radiation or a "plasma vortex." Some circle enthusiasts report hearing

strange noises and hums while standing in a circle as well as other spiritual experiences. Metaphysical connections have also been made between the circles and Stone Age sites nearby. Further, the circles are claimed to conform to "Sacred Geometry," the medieval idea that religious paintings and architecture should mimic the divine design of nature with spiritually enhanced mathematical constructions meant to transcend time and space.

The phenomenological aspect of crop circles took a hit in 1991 when two elderly British men, Doug Bower and Dave Chorley, admitted they had made the crop circles as a prank to spoof UFO believers. They said that they made the circles using a simple contraption of a board and some rope. Later, other hoaxers began to make circles in increasing complexity, though still using the tried and true method designed by Doug and Dave. Cerealogists (derisively referred to as "croppies" by critics) continue to argue that not all the circles are hoaxed. They point to the number of circles that appeared in the 1970s and 1980s before the wide public interest

Crop circles are rarely simple circles anymore. Now they appear in ever-increasing complexity.

began and noted that these could not be accounted for by the hokum of a couple of elderly pranksters. The patterns appear, they say, along magnetic earth lines, near Stone Age sites and that the plant stalk nodes are exploded out under strange conditions not caused by being crushed down by wooden boards.

See also: Unidentified Flying Objects (UFOs).

Further Reading

Hempstead, Martin. 1991. The summer of '91. *Skeptic* 5 (6).
Nickel, Joe. 1992. "The crop circle phenomenon: An investigative report." *The Skeptical Inquirer* (Winter): 136–49.

CRYPTOZOOLOGY

Cryptozoology is the search for "hidden" animals, those once thought to have become extinct but that still survive, animals that were thought to have been only mythical but which actually existed or were based on actual animals, and animals that have yet to be named by science. This field is one which teeters on the edge of legitimacy, as there have been examples of animals discovered that fit the criteria just listed: The most famous examples are those of the coelacanth, the okapi, the giant panda, and the mountain gorilla. The first, the coelacanth, is a lobe-finned fish thought to have become extinct along with the dinosaurs; the others were thought to be mythical by Western scientists who discounted stories of the creatures from local inhabitants. Unknown, noncryptid species are found by mainstream science on a regular basis.

Cryptozoology finds its modern genesis in the work of Bernard Heuvelmans (1916–2001) and Ivan Sanderson (1911–1973). Both trained scientists, they began writing about anomalous animals in the 1940s and 1950s. In 1955, Heuvelmans, a French-born Belgian zoologist, released *Sur la Piste des Bêtes Ignorées* (*On the Track of Unknown Animals*). This was the first influential work in the field and inspired legions of followers and intellectual children. He authored many books on the subject, including the seminal *In the Wake of the Sea Serpents* (1965). His friend and colleague, Scottish zoologist Ivan Sanderson, produced *Abominable Snowmen: Legend Come to Life* (1961). It was the first major work on what Sanderson called ABSMs (Abominable Snowmen). Heuvelmans attributes the term cryptozoology to Sanderson. In his work, Heuvelmans argues that the search for cryptids (individual organisms that fall into the category) must be thorough and scientific, as the object is to look for not only animals in the field, but also to discover the folkloric nature of such creatures. It is up to the cryptozoologist to weed through the mountains of legends, artwork, and national and cultural legends to find whether such stories and descriptions of strange animals have any basis in reality. Following this tradition, Adrienne Mayor produced the groundbreaking work *The First Fossil Hunters* (2000), which argued that many Greek legends of monsters, heroes, and other exotic life were attempts by ancient writers to make sense of their discovery of fossil bones. While Heuvelmans was a generalist, most modern cryptozoologists specialize in one particular area, as Sanderson did. Like Mayor, Richard Ellis, and Jeff Meldrum, they study fossils, sea monsters, or Bigfoot, respectively. Current leading practitioners who take a more generalist approach include Karl Shuker and Loren Coleman.

As an intellectual endeavor, cryptozoology has been studied as much as cryptozoologists have sought hidden animals. For example, literary scholar Peter Dendle has looked at the hunt for cryptids as filling a number of psychological roles—including as a vehicle by which to explore the problems of deforestation, environmentalism, and the profound impact of human society on animal species worldwide. In addition, he sees the fascination with cryptids as a holdover from medieval times and the wide array of modern books on the subject as the latest incarnation of illuminated bestiaries. Cryptozoologists focus on the search for large, imposing beasts that could be labeled monsters. They search for the dark, slouching hulks that lurk in the darker corners of the human psyche as much as they do the dark corners of the world.

One of the difficulties faced by cryptozoologists is explaining just what they do and how they go about it. Where zoologists, biologists, botanists, and the like have established methodologies for studying living things and general paradigms for making sense of what they are doing, cryptozoologists as yet do not. Though many practitioners attempt to be as "scientific" as they can, there are no accepted, uniform, or successful methods of pursuing cryptids. Certain technologies, like motion sensitive cameras, night vision devices, and audio recorders are utilized. Adding that none of the most famous cryptids, the Loch Ness Monster, Sasquatch, Yeti, and Chupacabra, has yet to be proven to exist with certainty, it has been difficult for cryptozoology to make a breakthrough into mainstream science. There have been attempts at codifying the methods of cryptozoology, thus legitimizing it, and individual amateur naturalists and professional scientists have explained how they went about researching this cryptid or that, but there are few uniform techniques articulated.

To date, the most coherent attempt to codify cryptozoological thinking has been Chad Arment's *Cryptozoology: Science and Speculation* (2004). He states that cryptozoology "is a targeted search methodology for zoological discovery" (9). The cryptozoologist looks for "ethnoknowns," life forms that have had human contact prior to scientific description through mythological and legendary appearances in written and oral tradition. Therefore, what separates a cryptid from the average unknown biological species, of which many examples exist, is the folkloric factor. If a biologist discovers a new species of beetle, plant, or bird—even if that biologist suspected it was there—it is not a cryptid. If the animal appeared in some culture's folklore prior to scientific discovery, then it is a cryptid. Therefore, Arment continues, the methodology a cryptozoologist follows is to collect mythical and legendary descriptions of a biological entity (he and most cryptozoologists are adamant that ghostly, apparitional, or otherwise paranormal subjects are outside the purview of what they do). They consult oral and written traditions, ancient texts, religious beliefs, and other sources for references to the entity. Using that as a basis, the researcher then compares those descriptions to known taxa to search for resemblances of behavior, morphology, and geographic location to posit possible matches. If the researcher thinks the organism extant, then a "targeted search" is made of the possible habitat the cryptid might occupy. If the organism is thought to be extinct, the museum collections where such fossils are found are consulted. While this is the general theoretical idea of searching for a cryptid, once researchers head out into the field, the idea of uniform methodologies has become somewhat more problematic. A number of techniques have been employed by both amateur naturalists and professional scientists in the field, but they have been applied unevenly.

See also: Anomalous Primates; Bigfoot; Jersey Devil.

Further Reading

Arment, Chad. 2004. *Cryptozoology: Science and speculation.* Lancaster County, PA: Coachwhip.

Dendle, Peter. 2006. Cryptozoology in the medieval and modern worlds. *Folklore* 117 (August): 190–206.

Heuvelmans, Bernard. 1958. *Sur la Piste des Bêtes Ignorées.* Paris: Plon. English trans. *On the track of unknown animals.* New York: Hill and Wang.

Sanderson, Ivan. 1961. *Abominable snowmen: Legend come to life.* Philadelphia, PA: Chilton.

CRYSTAL HEALING

An alternative medical practice in which crystals of various sizes, colors, and species are placed on or about the human body with the belief that this will have a beneficial therapeutic effect. It is believed that crystals focus and direct energy to the human body, helping to balance it and bring it in tune with the vibrations of the universe. Crystal healing became of interest to New Age purveyors because of the writings of the early 20th-century American psychic and prophet Edgar Cayce (1877–1945). He claimed that the denizens of Atlantis used crystals for healing and motive power. Crystal healing therapy should not be confused with the crystal gazing or scrying of prophets and those

seeking an image of the future. Healing crystals are placed on the body during therapy sessions or worn for long periods of time as talismans.

Crystals are used in science and industry in various roles. Cut and polished to certain shapes and standards, crystals become lenses for telescopes and microscopes, they amplify electromagnetic waves for radio reception, and are used in studying the structure of atoms. The structure of DNA was discovered using crystals. While many alternative healing aficionados claim crystals do have therapeutic powers, to date there is no proof that they do, other than from anecdotal eyewitness reports.

See also: Alternative Medicine; Crystal Skull.

CRYSTAL SKULL

A life-sized, human skull-shaped crystal carving said to have magical powers and portend evil doings. There actually are several "crystal skulls" said to conform to this description. They are located around the world in private and public collections and are of varying degrees of aesthetic quality. The category of anomalous crystal skulls are those that are in the life-sized range and are not of modern (i.e., late-20th- or early-21st-century) origins. The iconic crystal skull is known as the Mitchell-Hedges skull. It is rendered in the most anatomically correct detail, and the skull and jaw are separate pieces. The name comes from the presumed "discoverer" of the skull, British adventurer Frederick Albert Mitchell-Hedges (1882–1959). Mitchell-Hedges's adopted daughter, Anna Marie, found the skull in 1924 while the pair were looking for the lost city of Atlantis in a part of British Honduras that is

The Mitchell-Hedges skull is thought by believers to possess magical powers. Recent examination has shown modern tool marks on its surface.

now in Belize, Central America. Though they did not find Atlantis, the Mitchell-Hedges did find an unknown Mayan ruin called Lubaantun. It was at this site that Anna Marie found the skull under some stonework. This story is likely apocryphal and only appears for the first time in a vague form in Mitchell-Hedges's autobiography *Dander My Ally* (1954). He may have purchased the skull at an auction in 1943.

Of the other skulls, the one of best anatomical quality is in the collection of the British Museum. The others vary in size and are mostly highly stylized renderings. They are all variously attributed to "ancient" civilizations of Central America. The problem is that crystal is notoriously difficult, if not impossible, to date geologically with any accuracy. Also, the origins of these skulls are questionable at best. Tests done on the Mitchell-Hedges skull in 1970 by technicians at Hewlett-Packard company could find no evidence of tool marks on the crystal. Skull enthusiasts have claimed to hear the skull vibrate, sing, and generate spectral images and hallucinations. Some use the skulls for alternative medical therapies and psychological experiments. They have also made connections between the Mitchell-Hedges skull and the supposed Mayan prophecy of doom associated with the year 2012.

In the end, what we have is a collection of interesting, human skull-shaped crystal carvings. The faithful think the skulls were produced by lost races using hidden knowledge. They are said to vibrate with the music of the spheres and be portals to other dimensions and provide healing effects, and supporters attribute mystical powers and portentous meaning to them. With no way to accurately date crystal and given the unconfirmed stories of their origins, in the end we have nothing more than interesting crystal carvings. In 2008 more sophisticated tests showed that the Mitchell-Hedges did have marks of modern tools making it no older than the late 19th century.

See also: Crystal Healing.

Further Reading

Garvin, Richard. 1973. *The crystal skull*. New York: Doubleday.

DAY-AGE THEORY

Popular creation science notion that the "days" of creation mentioned in the Bible were of an indeterminate length greater than 24 hours. This allows for science-conscious Christians to reconcile the disparity of the Genesis account, which suggests a young earth, and the extensive geological evidence pointing to an old earth. Using this approach, one can argue that each day of the creation story might have been millions, if not billions, of years long. This would allow adherents to accept both a physical and metaphysical explanation for the age of the earth without contradiction. Following the Scopes Monkey Trial of 1925, after which Christian fundamentalism was thought to have been defeated, some anti-evolutionists moved away from a purely scriptural point of view and began building what they thought would be a scientific refutation of human evolution. Creationism is the belief that the earth and all life on it are the result of divine intervention. Creation science was a fundamentalist attempt to generate empirical evidence supporting a literal interpretation of the Genesis story of a six-day creation week occurring no more than 10,000 years ago that could be proven by scientific methods. Not all creationists took the Young Earth position. Some allowed that the earth might be billions of years old and used the Day-Age Theory as a way to support that idea. Strict literalists tend to reject the Day-Age Theory because it does not fit biblical accounts and thus contradicts God's revealed word. A related idea is the Gap Theory.

See also: Creation Science; Gap Theory.

Further Reading

Numbers, Ronald. 1992. *The creationists*. Berkeley: University of California Press.

Regal, Brian. 2005. *Human evolution: A guide to the debates*. Santa Barbara, CA: ABC-CLIO.

DEBUNKERS

Those skeptics who attempt to show the pseudoscientific and hoaxed nature of paranormal activities and beliefs. Debunkers tend to focus on spirit-related topics like ghosts, faith healing, spiritualism, claimed inexplicable human mind powers, and similar areas. Debunkers generally argue that what looks like a supernatural power or phenomena are, in reality, naturally occurring phenomena, tricks, misidentifications, and scams. The character of the debunker often takes the form of former magicians, psychologists, and atheists; debunkers are sometimes all three, but not always. Magicians tend to look for the scam angle, looking for how the trick is done, academic psychologists look to the psychological connections, and atheists argue that the paranormal cannot exist because the spirit world is an illusion. They all attempt to employ rational thinking and modern scientific technology and methodology to their efforts. They are often successful.

The term "debunk" can be traced to early-19th-century America. Congressman Felix Walker (1753–1828), who had fought in the American Revolution, was giving a long-winded speech in 1820 concerning one of the districts he represented in his home state of North Carolina, despite the fact that the issue at hand had little to do with his constituents. So tedious was his oration that fellow congressmen asked him to stop. He refused, saying he needed to assiduously represent the interests of the county of Buncombe. Wags in Congress began referring to any long-winded speech as "Bunkum" (in a misspelling). Eventually the word was shortened to bunk, and came to mean any empty political blather and then any idea thought to be nonsense. The word "debunk" first appeared in the written record in 1923 and meant to remove the preposterousness and obvious fakery of any event, idea, or belief system.

Investigations of paranormal activity have a long history. They became prevalent during the 19th-century Spiritualist movement and later Occult revival. These individuals are better termed investigators rather than debunkers because many of them believed psychic ability to be genuine and were searching for scientific proof of it. They also tended to be formal, serious, and scholarly. A genuine debunker may be thought of as one who does not believe in paranormal activity and attempts to prove its falsity. One of the first headline-grabbing investigators and genuine debunkers was the world-famous magician and escape artist Harry Houdini (1874–1926). Following the death of his beloved mother in the 1920s, Houdini became obsessed with contacting her in the afterlife. He consulted a number of spirit mediums and engaged in séances. To his great disappointment, he discovered the mediums were frauds and charlatans and set about on a public campaign to expose them. He was able to show how theatrical effects, stage lighting, and hidden assistants, just as he had used in his own stage act, could be employed to fool an audience into thinking a genuine contact had been made with the departed. He set the model for the magician/stage performer model of paranormal investigator that would later include James Randi, Joe Nickell, Penn and Teller, and others.

The late-20th-century equivalent to Houdini as debunker is James Randi (though Randi prefers to be called an investigator of the paranormal rather than a debunker). Also known as "The Amazing Randi," he began his public life as a professional stage magician (he has no formal training as a scientist or engineer). In the early 1970s, he became outraged by how some unscrupulous magicians were using their talents not to just entertain an audience, but to suggest that they had supernatural powers. Such suggestions of

paranormal ability go back to the earliest days of conjuring and were nothing new to the 20th century. A young Israeli named Uri Geller who claimed to be able to bend metal using the power of his mind especially bothered Randi. Randi called Geller a fraud and a series of legal suits for defamation ensued. Randi became a fixture on television explaining how various paranormal charlatans and hucksters went about their business. He has spoken out about faith healers like the Brazilian John of God, Filipino psychic surgeons, and American televangelists. In 1980, Randi was responsible for exposing the martial arts mystic with supposed telekinetic powers, James Hydrick, on television. Also in the 1980s, his investigation of Peter Popoff, a popular and wealthy Christian fundamentalist, faith healer, and television personality, showed that Popoff was wearing a hidden microphone that allowed his wife to feed him information about the names and medical conditions of church service attendees whom he "cured" live in front of huge audiences. Randi showed that Popoff was a charlatan, bilking hapless sufferers, most of whom were desperately poor and could not afford it, out of huge sums of money. The Popoff ministries soon failed in bankruptcy. By the early 21st century, however, Popoff had begun a return. He is now airing infomercials around the globe selling "Miracle Water." It is common for such "Holy Men" to be exposed as frauds and thieves but to then be forgiven by their followers who apparently know no end of gullibility.

Joe Nickell is another debunker. He calls himself "the world's only full-time professional paranormal investigator." Writing for the magazine *The Skeptical Inquirer*, Nickell has investigated ghosts, monsters, the Shroud of Turin, and other similar phenomena, and has found them all lacking. Also a former stage performer, he holds a doctorate in literary studies focusing on folklore. Michael Shermer's work is in a similar vein. An historian of science, founder of the Skeptics Society and editor of *Skeptic* magazine, Shermer has made a career out of exposing pseudoscience and paranormal fraud. In addition to the usual debunker topics, he has also investigated Holocaust deniers. In *Denying History: Who Says the Holocaust Never Happened and Why Do They Say It?* (2002), Shermer rightly argues that the best way to deal with Holocaust denial is not to suppress it, but to bring it out into the light of day, deconstruct its underlining assumptions, and show its inherent absurdity and mean-spiritedness.

Science, at its core, is a skeptical endeavor. Scientists and historians are always on the lookout for dubious claims, assertions that have no supporting evidence, or ideas and behaviors that resist examination. There is a saying, popular in skeptical circles, that extraordinary claims require extraordinary evidence. The more outlandish a supposition, the greater the need to prove it accurate. The pseudoscience investigator should always demand to see the proof of something. This is the underlining concept of debunking.

See also: Ghost Hunters.

Further Reading

Kalush, William, and Larry Sloman. 2007. *The secret life of Houdini: The making of America's first superhero.* New York: Atria.

Nickell. Joe. 2007. *Adventures in paranormal investigation.* Lexington: University Press of Kentucky.

Randi, James. 1982. *Flim-flam! Psychics, ESP, unicorns, and other delusions.* New York: Prometheus Books.

Shermer, Michael. 1997. *Why people believe weird things: Pseudoscience, superstition, and other confusions of our time.* New York: Holt.

DIVINATION

The art of fortune-telling through the use of various occult techniques and artifacts, divination is similar to prophecy in that it crosses all cultural and ethnic borders and goes back to the ancient world and beyond. The process has a questioner ask a diviner about the future. These are often personal questions about how to proceed with a transaction, what sort of decisions should be made, and what the future holds for them, but they can also be about other individuals or the future in general. The diviner then will scatter objects like pebbles or sand (geomancy), shuffle and then turn over cards in a prescribed method (Tarot), or allow another agent, like the wind, ripples on water, or even animals to move the objects around. Almost anything that creates a pattern can be used for divination, including cloud formations and even the flight of birds. It is the pattern that is looked for. The diviner then "reads" the objects through their patterns to guide him in formulating an answer. The diviner can respond with simple yes or no answers, more complex or straightforward ones, or ones couched in metaphor or riddle. Divination can also be done by reading large-scale events and patterns in human behavior. A popular form of divination is scrying, in which a diviner uses on object like a crystal ball, cup of water, or mirror to generate spiritual visions of events. While any object can be used for scrying, reflective surfaces are best and crystals preferred. The scrying object, sometimes called a shewstone, is used as a point for the diviner to focus her concentration on. In this way she enters into a kind of trance state in which she can see visions in the crystal.

The most famous ancient form of divination device is the Chinese *I Ching* (*Book of Changes*). Thought to have originated with a quasi-legendary Chinese mystic named Fu Xi (a kind of Chinese Hermes Trismegistus) in the early 2700s BC, the *I Ching* is a combination work of fortune-telling and deeper cultural philosophy. A set of 64 groups of lines were eventually codified, arranged, and can be read, and each set of lines has specific meaning and portents. A famous and influential early modern diviner was the English mathematician, astronomer and occultist John Dee (1527–1608). Dee initially used scrying to divine his dreams via a small crystal ball set in an elaborate mounting on a special table. He later employed a shewstone and a small Aztec ritual object he obtained and called a speculum. Another great scryer and diviner was Nostradamus (1503–1566), who may have used a pan of water for his scrying.

See also: Geomancy; Numerology; Second Sight.

Further Reading

Loewe, Michael, and Carmen Blacke, eds. 1981. *Oracles and divination*. New York: Shambhala/Random House.

Place, Robert. 2005. *The Tarot: History, symbolism, and divination*. New York: Tarcher.

rod, dips down to indicate a hit. Late-20th- and early-21st-century dowsers discarded the Y-shaped branch in favor of specially made separate metal rods, one held in each hand. Instead of the rods dipping down upon locating the target, they cross. Dowsing, in some form, goes back to antiquity, but dowsing as it is practiced today finds its origins in Germany in the 1400s, when it was used as a technique for finding mineral ore deposits. Both Catholic Church officials and leaders of the Protestant reformation denounced it as witchcraft and devil worship. A pendulum can also serve as a kind of low-tech metal and energy detector. The dowser holds a pendulum over an object, picture, map, or writing and, in a form of divination, taps into the earth's energies to locate a point on the object. Today, ghost hunters and other paranormal investigators use dowsing to locate spirits. There have been one or two attempts to study dowsing scientifically, but the results are inconclusive. As with so much of the paranormal, there is little evidence of the success of dowsing that is not anecdotal.

DOWSING

A technique for finding underground water sources, mineral deposits, or other objects through occult means, dowsing is an ancient process where the dowser carefully holds a Y-shaped branch or stick and feels for vibrations emanating from the ground. When the dowser is over the sought-after target, the branch, or dowsing

Dowsing in this way with a Y-shaped branch is a traditional folk method of finding hidden water sources.

See also: Divination.

Further Reading

Lonegren, Sig. 2007. *Spiritual dowsing: Tools for exploring the intangible realms*. Glastonbury, UK: Gothic Image Publications.

Webster, Richard. 2003. *Dowsing for beginners: How to find water, wealth and lost objects*. Woodbury, MN: Llewellyn Publications, 2003.

EASTER ISLAND

Remote Pacific Ocean island on which large carved volcanic stone heads are found. Anomalists have argued these heads could not have been constructed or carved by the culture known to have inhabited the island, and so argue that some extra-terrestrial assistance was used or that the Easter Islanders had access to forgotten knowledge. Known to the native people as *Rapa Nui,* Easter Island received its popular name in 1722 when Dutch explorer Jacob Rogeveen landed there on Easter Sunday. The first inhabitants may have arrived on the island as early as the 3rd century AD. There are pictographs of fish and bird-human hybrids there, however, the island is best known for the plethora of large human heads, called *Moai,* carved out of volcanic rock that dot the island. Stylized rather than realistic, the *Moai* are indeed some of the most interesting and enigmatic of all the carved stone monuments on earth. There are 394 surviving *Moai,* out of a total population thought to have been over 880. The large size of the *Moai,* many of which stand upon large

The Moai on Easter Island stare enigmatically off into space.

stone pedestals and altar-like structures, how they were carved and transported to their final positions, the small human population (even at the height of Rapanui culture), the lightly forested condition of the island, and its extreme isolation have added to the mystique of the place. The island has inspired some paranormalists to suggest alternative explanations for how the statues were erected. The Rapanui themselves refer to using a spiritual power called *Mana* to explain how they moved the multi-ton statues. This has led some to say the Rapanui had special knowledge of antigravity and other "earth forces." UFO author Eric Vön Daniken suggested in *Chariots of the Gods* that space aliens either erected the statues or assisted the Rapanui to do it. Extensive archaeological investigations of the island have shown a good bit of detail about the history of the Rapanui people: There was a series of civil wars, massacres, and over forestation. This has helped quiet some of the pseudoscientific speculation. The reality of Easter Island's history shows that it can be seen as a cautionary tale about the effects of human agency on the environment and the perils of ecological disaster. The island is now protected as a World Heritage Site and is a possession of Chile.

See also: Pseudohistory; Von Däniken, Eric.

Further Reading

Lee, Georgia. 1992. *The rock art of Easter Island: Symbols of power, prayers to the gods.* Los Angeles: The Institute of Archaeology Publications.

ECTOPLASM

Diaphanous material said to be produced by spirit mediums during trances. During the 19th-century séance craze, mediums regularly produced ectoplasm as part of their performance. Ectoplasm in the form of a usually milky, paste-like substance has been reported to take the shape of objects like trumpets and human body parts, such as limbs and faces. Just what ectoplasm is made of is difficult to say. French physiologist Charles Richet (1850–1935) in his investigation of the phenomena coined the term ectoplasm in 1913. He called it the "exteriorized substance" emitted by the medium during a trance. Richet was awarded a Nobel Prize for medicine for his work on anaphylaxis, but was also interested in the paranormal. Ectoplasm has appeared much less often in the early 21st century. It does not help the case of paranormal believers that most, if not all, mediums who produced ectoplasm were shown to be frauds and the few times samples were collected for analysis they proved to be of mundane materials.

See also: Ghost Hunting.

EMPIRICISM

Any knowledge gained through firsthand, direct observation of phenomena. Empiricism is fact-based knowledge rather than something learned from secondary sources. A form of epistemology (the study of the nature of knowledge), empirical data is at the heart of the scientific method. Carefully collected and collated data which supports a hypothesis, premise, or theory can thus be double-checked by outside parties for accuracy and efficacy. To know something empirically is considered one of the most reliable ways to know. In science, empirical data is gathered through experimentation, observation, and other direct experiential methods.

Further Reading

Richardon, Alan, and Thomas Uebel, eds. 2007. *The Cambridge companion to logical empiricism*. Cambridge: Cambridge University Press.

Robinson, Dave. 2004. *Introducing empiricism*. Blue Ridge Summit, PA: Totem Books.

EUGENICS

The idea that a carefully controlled program of human breeding can improve society. First appearing in the 19th century in the writings of British social theorist and statistician Francis Galton (1822–1911), eugenics became popular in the United States, South America, and Scandinavia before spreading quickly around the world and reaching its peak in the extermination program of the Nazis in the 1930s and 1940s. Eugenic belief presupposes that there are superior and inferior ethnic groups; it also holds that both the physical and metaphysical characteristics of an individual or "race" are determined by the quality and character of their ancestry. Whatever characteristics a person or group is born with are predetermined and permanent: no amount of social amelioration can change a person's circumstances. Following this logic, eugenicists argued that to "clean up" a society, it had to be rid of its tainted blood. This was to be done by first restricting outside immigration, segregating populations, sterilizing the unwanted, and ultimately eliminating them. Once thought of as "good" science, the tenets of eugenics have since been shown to be baseless and without merit from either a scientific or humanistic point of view.

Eugenics has its roots in a very common concept; the idea of arranged marriages goes back to antiquity. Royal houses

Madison Grant was the most notorious American eugenicist. He inspired many others, including Adolf Hitler.

and religious groups regularly decided who could or could not marry. Eugenics was the 19th-century version of this idea, but it connected the authority of science to it. Discoveries about the nature of heredity and the reality of evolution (Galton was Charles Darwin's cousin) helped give a supposed authority to previously held beliefs. Galton was a pioneer in the application of statistical methods to the social sciences. He determined that successful middle and upper-middle-class people were being outbred by a rising tide of lower-class "degenerates" and that something had to be done or the British Empire would be overrun and ruined. His original ideas incorporated soft or "positive eugenics." This was the idea of the state providing financial rewards to encourage the "right sorts" to procreate more and the "'lower orders" or "social residuum" to breed less. Eugenicists believed that personal traits like courage, poverty, illiteracy, business success, alcoholism, honesty, and criminality were inherited traits.

The idea of racial inequality goes back as far as human civilization itself, but was codified into its modern form in the work of Frenchman Joseph-Arthur de Gobineau who, in the 1860s, wrote *The Inequality of the Races.* Gobineau argued that human society was a struggle between superior and inferior races; he was both antidemocratic and anti-Semitic. Gobineau and other like-minded individuals saw racial inequality as a staple of the human experience. To them, Caucasians were biologically, intellectually, and culturally superior to other ethnic groups. Within the wide range of what were believed to be Caucasian types— Nordics, Mediterraneans, and Alpines— the greatest were the Nordics who could trace their lineage back to the mythical Aryans. They were the best form of human evolution, the cream of the evolutionary crop.

Eugenics took its first practical form in the United States in the late 19th and early 20th centuries. Nativists—those born in the United States of Nordic, Anglo-Saxon ancestry—were deathly afraid of the large number of new immigrants. The eugenicists felt the same way but brought to the discussion what they saw as a practical remedy. Beginning in the 1890s, the nativists and eugenicists claimed they would apply "scientific" techniques to cleansing via sterilization. The eugenicists added to the basic nativist fear of foreign culture a fear of foreign biology in the form of what they called "feeblemindedness." Feeblemindedness was a catch-all expression meant to describe mental disabilities. Its meaning was soon made so elastic that it could be used to cover almost any social ill the eugenicists wanted it to, from severe mental impairment to simple illiteracy or poverty. In addition, miscegenation, or race mixing, was raised to the level of biological warfare. Because eugenicists believed blacks to be of a different and inferior evolutionary origin, they wanted the races kept apart. Any such mixing, they argued, would throw off the delicate balance of human evolution.

Impetus was given the eugenics movement by the rediscovery in 1900 of mid-19th-century Hungarian monk Gregor Mendel's (1822–1884) laws of heredity. He had done extensive work on botanical heredity and learned how to manipulate the color of pea plants. This suggested that physical traits were the result of "unit characters"—single hereditary units that were responsible for individual characteristics like color and height. Mendel's work was published in an obscure local journal and did not become known to the wider scientific community until the end of the century. Many researchers interested in heredity took Mendel's law to mean that if a particular trait was unwanted, it could be eliminated from the population. Confusing social with biological traits became a cornerstone of eugenic thought; to clean up a population biologically, they argued, one must identify unwanted traits and then eliminate those individuals who carried them, or keep them from procreating or mixing their blood with those who did not have the traits.

Along with Mendel's work, eugenicists also adopted the germ-plasm theory of German cytologist (one who studies cells) August Weismann (1834–1914). In the 1880s, Weismann discovered that the human body contained two types of cells: somatic and reproductive. He showed that only hereditary material contained in reproductive cells was used to create the next generation. Eugenicists fixed on this as proof that it was blood that carried superior and inferior traits, making the mixing of bloodlines harmful. They saw this information as evidence that biology alone (nature), not environment and upbringing (nurture), was responsible for transmitting traits through the bloodstream.

Possibly the most notorious purveyor of eugenics in the United States was Madison Grant (1865–1937). A wealthy New York lawyer, Grant argued in *The Passing of the Great Race* (1916) and *The Conquest of a Continent* (1933) that the country was being swallowed up by hoards of degenerates whose inferior origins were destroying the fabric of society. Grant used as an underlying framework the central-Asia hypothesis of human origins put forth by his friend Henry Fairfield Osborn who claimed Aryans evolved in central Asia under separate and different origins than the other races before they swept across Europe, becoming Nordics and Anglo-Saxons, to create great civilizations that found its highest expression in North America.

Grant and other prominent American eugenicists lobbied the U.S. government to enact stricter immigration reforms to stem the tide of unwanted masses they feared were entering the country, polluting Anglo-Saxon blood, and ruining Anglo-Saxon society. The result was the Johnson/Reed Act of 1924 that restricted the numbers of Asians, Irish, Slavs, southern Italians, and other groups that Grant and his cronies both in and out of the government deemed undesirable. This was the first permanent immigration law in the United States and remained in force until 1952. Grant also put forward a three-tiered plan

for dealing with this evolutionary threat. The first was to restrict new immigrants from coming into the country. Second, any undesirables already in the country should be segregated into special areas so that they could not mix with the Anglo-Saxon population. Finally, all undesirables should be sterilized, by force if necessary, to ensure they would die without issue. Sterilization was a favored remedy of eugenicists for fixing all the country's ills.

The pivotal moment in the sterilization debate in the United States was the *Buck v. Bell* case. Seventeen-year-old Carrie Buck was an institutionalized woman from Virginia. Deemed "feebleminded" because she and her mother, Emma, became pregnant out of wedlock, Carrie was placed in the Virginia Colony for Epileptics and the Feebleminded in 1927. The state of Virginia decided to sterilize her and her child, Vivian, so they would not reproduce again. A legal action ensued that went to the Supreme Court. The state argued that it had the right to sterilize any citizen posing a threat to the state's economy through reproduction.

The Eugenics Record Office (ERO), a biological laboratory on Long Island, New York, came to the state's defense. The ERO was started with money from the influential Harriman family and led by biologist Charles Davenport (1866–1944). It was a center of eugenic research focusing on how to control and modify the American breeding public (Davenport was a friend and ally of Madison Grant). The ERO collected data on thousands of Americans in an attempt to chart the country's hereditary disposition. These data would then be used to construct a plan to control breeding; this meant finding ways to keep "degenerates" from procreating and to promulgate the births of the "right sorts."

To aid the state of Virginia, Davenport sent Dr. Arthur Eastabrook to testify. After a brief exam of Carrie's daughter, Vivian, he declared her "backward" (she was, in fact, a straight-A student). The head of the colony, Dr. Albert Priddy, testified that the entire Buck family was a tangled bush of degenerates badly in need of some pruning. The defense argued that the state had no right to sterilize anyone against their will. In the end, the Supreme Court sided with Virginia and allowed the procedure to take place. In his majority statement, Chief Justice Oliver Wendell Holmes Jr. said the state had the right to perform such operations to protect itself. He argued that if too many mentally handicapped or degenerate people were allowed to breed, they would adversely impact the entire nation. The *Buck v. Bell* case was a watershed for sterilization in the United States. Almost immediately, states across the Union began passing sterilization laws that resulted in the forced sterilizations of tens of thousands of Americans. Eugenic sterilization programs also began to appear around the world, in Latin America, Europe, and Asia.

With their rise to power in the 1930s, the Nazis of Germany were also eugenicists and wanted to rid their country of all undesirables. They patterned their race and sterilization laws on those of the United States. The Nazis, with the help of the German medical and scientific community, began by euthanizing ("mercy" killing) mentally handicapped and physically deformed individuals in their hospital system. This program was later used as a model for the Nazis major eugenics program, the extermination of all European Jews, and came to be called the Holocaust.

Further Reading

Kevles, Daniel. 1998. *In the name of eugenics.* Cambridge, MA: Harvard University Press.

EVPs

Supposed sounds of ghostly entities recorded at the site of a haunting. Electronic Voice Patterns (EVPs) are those sounds not heard by human ears at the moment the recordings are made, rather, they are only heard later when the recording device is played back. EVPs are often indistinct, though they occasionally are clear and sound like human voices speaking intelligible sounds. Ghost hunters bring audio recording devices to the sites they investigate in hopes of capturing EVPs. One popular technique for eliciting EVPs is for investigators to ask questions of the entities they seek. Later, the recording devices are played back and EVPs are searched for. They often come in the form of short phrases and one- or two-word utterances. EVPs are different than spiritualist communications, as they are far more random and haphazard than formal séances. The problem with EVPs is that little research has been done into their nature. No protocols have been worked out for how to procure them or what to do with them once they are found. A case can be made for the idea that many EVPs are artifacts of the recording process itself with which the operators are unfamiliar. The majority of EVPs have alternative, nonspiritual sources; anomalous ones have no clear proof they are of spiritual origin.

See also: Ghost Hunters.

F–K

FACE ON MARS

Surface details in the Cydonia region of the planet Mars, which some interpret as the remains of designed structures meant to represent humanoid faces, pyramids, and other similar artifacts thus implying that some nonhuman intelligent agent constructed them. If this interpretation is correct, it would prove the existence of sentient beings outside of the earth. Scientists at the United States' National Aeronautics and Space Administration (NASA) claim the image is a simulacrum of a face caused by sunlight striking naturally formed geologic features in just the right way. The photos were originally taken by the Viking spacecraft as it orbited Mars in July 1976. Eighteen pictures were taken of the Cydonia region. Analysis of the pictures revealed that one photo showed a raised feature that looked uncannily like a stylized human face and several nearby that resembled Egyptian-style pyramids. Two more photos of the same region showed the same feature with the face. The pictures were released to the public in 1977 and were immediately controversial.

The most persistent supporter of the notion that the Cydonia features are not simulacra, but designed structures, is science writer and former television space consultant Richard Hoagland. He is also a proponent of his own esoteric concept of multiple planes of existence called hyperdimensional physics. Hoagland argues that the features seen on Mars are the remains of a once-mighty civilization that constructed the mounds and pyramids so that humans would eventually see them and know that their civilization existed. He also suggests that, as their world was dying, the Martians moved to the earth and performed genetic engineering upon humans. Like the faked moon landing proponents, Hoagland argues that the reason the pictures of the face were released to the public—when logically, if NASA was covering up, they would have made their own lives easier by just keeping the photos secret—were through the effort of unknown, but courageous whistle-blowers, who, while working on the inside, managed

This raised structure on the surface of Mars is interpreted by some as evidence of intelligent extraterrestrial life. NASA scientists say it is a trick of the light playing upon a natural rock formation.

to slip the most important pictures ever taken past government censors. Others have picked up on Hoagland's work and claimed that the structures are said to conform to sacred geometry and are related to, variously, the ancient Sumerians, Egyptians, Hopi Indians, and Aztecs.

Hoagland also contends that photographs taken on the moon by Apollo astronauts show the remains of a lunar civilization. He has examined the photos in close detail and claims that, unlike the built-up earthen structures of Mars, the lunar photos show towers, spires, support cables, and other architectural and fabricated structures of glass and steel—or at least some lunar civilization equivalent. He interprets one photo as showing the broken and discarded head of a robot ly-

ing in a crater. Like the Martian photos, the lunar pictures are of poor resolution and the objects in them are seen at a distance. As for the men who actually walked on the moon and who could verify Hoagland's assertions, but do not, he puts their stories down to brainwashing by the government so they would not remember. He has made reference to Masonic conspiracies and, along with the fake moon-landing movement, suggested the U.S. government intentionally killed the crew of Apollo 1.

By the 1990s, other spacecraft, including the Global Surveyor, Reconnaissance Orbiter, and European Space Agency's Mars Express, had returned to Mars and photographed it with improved camera equipment. Prompted by Hoagland's and other Mars face aficionados, NASA had the ships make a special pass back over the Cydonia region to rephotograph the face and pyramid features. In 1998, those photos were released and showed even more that the features of the face and the pyramids were simply geologic formations. The face supporters now claimed that NASA was still covering up the truth and that the photos had been retouched. The face supporters, of course, were able to get the original photos without manipulation and showed how clear the face and pyramids were. Face proponents now claimed that new structures could be seen, including a "glass tunnel" and even a Martian "flying saucer" parked on the edge of a mountain ridge.

Planetary scientists and astronomers alike contend that the moon has never been able to support any life and that while Mars may have had some liquid water and some of the primary building blocks of life as humans know it, it never had and any life forms and certainly no sentient beings who built vast networks of structures. Hoagland and his supporters wave all this off as cover-ups, blinkered thinking, and intentional wrongdoing. NASA scientists who try to explain what the "face," "pyramids," and other supposed structures on Mars and the Moon are fight an uphill battle. The more they show what these features are, the more they are accused of lying about it.

Taken together, the Martian face, flat earth, hollow earth, and moon hoax advocates, while seemingly at odds with each other, do have a common thread to their work. They all argue that there is a nefarious government cover-up meant to keep the "truth" about this hidden history from being known. This is also a common thread to pseudoscience in general. The reason their work is not taken seriously by science, they claim, is because an ill-defined force is holding them down, stifling descent, and generally behaving badly. If only the force were lifted, they say, these seekers of the truth would be held up as heroes and pioneering geniuses.

See also: Flat Earth; Hidden History; Hollow Earth; Moon Landing Hoax.

Further Reading

Hoagland, Richard. 2002. *The monuments of Mars: A city on the edge of forever*. Mumbai, India: Frog Books.

Hoagland, Richard, and M. Bara. 2007. *Dark mission: The secret history of NASA*. Port Townsend, WA: Feral House.

FAIRY PICTURES

Photographs of alleged fairies taken by British cousins Elsie Wright and Francis Griffiths between 1917 and 1920. Fairies are mythical, spirit beings, whose exact nature is disputed even by believers. The word normally conjures up images of delicate female entities, but it can also be used to describe less attractive ones, like goblins. Elsie borrowed her father's camera in July 1917 to take pictures in the woods around their home in Cottingly, England. When he developed the film he saw fairies in one picture. He thought the girls were playing a prank, though his wife thought the picture genuine. Another picture was produced in September of that year. The first picture showed Francis with a group of cavorting fairies; the second was of Elsie and a single fairy. In 1919, Mrs. Wright, who had an interest in the occult, attended a Theosophical Society meeting and told fellow attendees about the pictures. Author Arthur Conon Doyle (1859–1930) was an ardent Spiritualist and was at the time working on an article about fairies. The creator of Sherlock Holmes, Doyle was famous and his work widely read. He included the two pictures in his article and they caused a sensation. Camps immediately appeared of those who supported the pictures and those who denounced them as obvious fakes. In 1920, interested parties gave the girls a camera and film in the hopes that more pictures would be forthcoming; they were: Three more pictures were taken, for a total of five. Doyle felt vindicated and wrote, *The Coming of the Fairies* (1922). By this time, the girls had grown tired of the game and stopped taking pictures. They slipped into quiet obscure lives until the 1980s. Beginning in 1981, the now-grown women gave a series of interviews in which they admitted they had faked the photos by copying images from a popular picture book, *Princess Mary's Gift Book* (1914), cutting them out, coloring them and attaching them to long hat pins to fix in place. They said they became scared when the media hysteria began and were afraid to admit that it was all a bit of little girl

The famous fairy pictures, which convinced so many adults of their reality, were paper cutouts made by a pair of mischievous girls.

fun that got out of hand. Enigmatically, though, at times they said all the pictures were fake, and occasionally said all were fake except the fifth photo. They have also claimed that while they did not photograph fairies, they did see them.

Further Reading

Gardner, Edward L. 1974. *Fairies: The Cottingly photographs and their sequel.* Quezon City, Philippines: Theosophical Publishing House.
Randi, James. 1982. *Flim-Flam!* New York: Prometheus Books.

FALSIFICATION

An important element in the definition of science, falsification is an idea put forward by Austrian economist Karl Popper (1902–1994). Popper rejected empiricism as the only criteria for science and argued that scientific theories are, by nature, abstract things that need more than simple facts. He argued that even an infinite number of positive tests and experiments cannot confirm a theory, but only one negative example is needed to show it to be false. However, falsification does not mean the idea is a false one, rather it only means there is the possibility that it could be proven false. If there is no chance to prove an idea false, it is not falsifiable. Popper also claimed that if an idea cannot be falsified but still can be used to answer a wide range of questions about a field of inquiry consistently and uniformly, it can be judged to be science. It is the idea of falsification that demarcates science from pseudoscience. For example, a photo of a UFO cannot prove UFOs exist. The photo itself can be tested, but the results only pertain to the physical photo itself. Whether or not the photo was a hoax does not prove or disprove the reality of UFOs. As a result, the belief that UFOs exist, that they come from other worlds, or that they have been visiting the earth is not a scientific theory, because, as of yet, it cannot be proven to be false.

See also: Paradigms.

Further Reading

Losee, John. 2005. *Theories on the scrap heap: Scientists and philosophers on the falsification, rejection, and replacement of theories.* Pittsburgh, PA: University of Pittsburgh Press.
Popper, Karl. 1959. *The logic of scientific discovery.* New York: Basic Books.
Popper, Karl. 1963. *Conjectures and refutations.* London: Routledge.

FLAMEL, NICHOLAS

French alchemist reported to be one of the few people — along with his wife Perenelle — to have successfully created the Philosopher's Stone and, as a result, large amounts of gold. Flamel (1330–1417) is also reputed to have lived, by use of the Philosopher's Stone, several centuries, having been seen in India and Paris as late as 1700.

Flamel was a professional scribe who made his living copying books and official documents. In middle age he had a dream in which he found a fabulous book on alchemy. He then found such a volume in a book-seller's stall. Amazed and intrigued by the coincidence, he and his wife spent years trying to understand the strange text without success. Frustrated, he went on a pilgrimage to Spain and encountered a Jewish scholar who told him the book was an ancient text on the Jewish mystical tradition known as Kabbalah. Flamel eventually returned home with an understanding of how to employ the book's directions. Early in 1382, he cracked the code, produced a Philosopher's Stone, and then copious amounts of gold which was soon being handed out to various worthy charities.

Contemporaries of the Flamels felt so strongly that they had indeed discovered how to produce gold that, following their deaths, their house was ransacked and their graves violated in search of their secrets. While Nicholas Flamel certainly lived, pursued alchemy, and was a philan-

Nicholas Flamel is one of the few people purported to have created a Philosopher's Stone.

thropist, that he discovered the Philosopher's Stone and lived until 1700 are romantic legends with at best circumstantial evidence supporting them. He did achieve a kind of immortality, however. His memory found resurgence outside of the historical community when he was used by novelist J. K. Rowling as a central, but unseen character in both the text and film version of *Harry Potter and the Philosopher's Stone* (1997).

See also: Alchemy; Philosopher's Stone.

FLAT EARTH

The idea that the earth is a flat plane rather than a sphere. Belief that the earth is flat goes back to humans' earliest attempts to determine its size and dimensions. There is a common misconception that, until the Age of Discovery, and particularly the voyages of Christopher Columbus, people believed the world was flat. This became common in North America after the publication of Washington Irving's novel *The Life and Voyages of Christopher Columbus* (1828). The idea was repeated in a number of American history school textbooks as fact, despite its fictional origins. The 19th-century historical contrivance of a European "Dark Age" also contributed to this inaccuracy. By the time of the ancient Greeks, however, it had been determined that the earth was round. Pythagoras and Plato both taught the earth was a sphere, as did Aristotle. In 240 BC, Eratosthenes determined the earth's circumference using measurements of the cast shadows of wooden stakes along with trigonometric calculations. His final answer was very close to modern scientific measurements. Indian and Islamic scholars had also determined the spherical nature of the earth. By the 20th century, techniques of geodesy (measuring the earth and studying its magnetic fields) and aerial and space photography had conclusively confirmed that the planet was round and what its size and weight was. Despite this evidence, however, a small number of people still believe the earth was flat.

A major proponent of modern flat earth theory was English inventor Samuel Rowbotham (1816–1884). Known by his pseudonym of "Parallax," Rowbotham developed a method for determining a flat earth he called Zetetic Astronomy. He developed a model of the flat earth that had the North Pole at the center of a large, flat disk. The South Pole was the perimeter of the disk, and the equator a circle midway between the center and edge. The sun, moon, and stars rotated above this plane forming a kind of dome over the top of the disk.

Rowbotham drew thousands of followers who after his death created the Zetetic Society. One of his most ardent followers was fellow Englishman William Carpenter (1830–1896) who moved to Baltimore to engage in his primary profession of printer. Carpenter produced *One Hundred Proofs the Earth Not a Globe* (1885) to further the idea. He printed tens of thousands of copies, with constant new editions, most of which he distributed free to libraries, religious groups, and scholars. Like other Flat Earthers, Carpenter pointed to rivers not dropping in their levels, large bodies of water not showing convexity, and arguments that, when a ship looks like it is disappearing over the horizon, it is an optical illusion and the result of complex perspective. Rowbotham's musings on a flat earth were of interest to Christian fundamentalists in North America and by one religious sect in particular, who called themselves the Christian Catholic Apostolic Church of Zion, Illinois. The Apostolic Church saw Zetetic Astronomy as "scientific" proof of their Biblically based belief that the earth was flat. Originally begun as a utopian community by faith healer John Alexander Dowie, Zion was taken over by Wilbur Glenn Voliva (1870–1942), a Flat Earther and aspiring megalomaniac. He made flat earth belief the central tenet of the community. In addition to a round earth, the other things Voliva didn't care for were science, evolution, tobacco, doctors, liquor, unions, or fun. He used biblical proofs to support the flat earth idea as well all the standard pseudoscientific, antiround-earth arguments.

In 1922, Voliva built a radio station and began some of the first religious broadcasting. He railed against evolution, biblical criticism, a round earth, and modern astronomy in general. In 1931, he managed to position himself as the focus of an article on flat earth belief in the magazine *Modern Mechanix and Invention*. He offered $5,000 to anyone who could prove the earth round. As with all such challenges — prove evolution, prove an old earth, and so on—Voliva had no intention of ever paying the reward. He had answers for any conceivable example and simply wanted to bait scientists and flat earth detractors into arguments from which they could never get disentangled. The article did include a series of charming illustrations of boats near the frozen edge of the disk and the sun and moon orbiting "above around" the earth. The Zion community eventually broke up due to Voliva's heavy-handedness and dictatorial ways. Furthermore, the state of Illinois had questions about his financial dealings.

The primary promoter of flat earth theory in the later 20th century was the Flat Earth Society. Founded in 1956 by yet another British Flat Earther named Samuel Shenton (d. 1971), it was inherited by American Charles K. Johnson (1924–2001) when Shenton passed away. Johnson also promoted the idea that the Apollo moon landings had been hoaxes and made claims about the hoaxed nature of the space shuttle program. He said that God had destroyed the shuttle *Challenger*, and argued in his society's newsletter that the moon was internally lit, not illuminated by the rays of the sun. The space program was a none-too-artful attempt to lead people away from Jesus by getting them to believe the earth round, in contradiction to scripture. A Christian fundamentalist, Johnson was contemptuous of fellow creation science proponents despite being as much a creationist and geocentrist as they were. Flat earth proponents also tend to be neogeocentrists: Not only do they believe the earth is flat, they also believe it is at the center of the universe. Johnson's publications were vituperative and angry, railing against science and skeptics. In one newsletter, Johnson ranted almost incoherently (like many conspiracy theorists, he loved using all capital letters for emphasis), equating science to grease and claiming that late American president Ronald Reagan was a secret Flat Earther because he looked upward to heaven during a speech. The Flat Earth Society descended from the British Zetetic Society, but Johnson claimed Moses originally founded it. Like the Zion community, the Flat Earth Society went into decline after its leader's demise and all but disappeared. There are online groups calling themselves the Flat Earth Society, but these are largely satirical.

Further Reading

Garwood, Christine. 2007. *Flat earth: The history of an infamous idea*. London: Pan Books.

Miller, Jay Earle. 1931. $5,000 for proving the earth a globe. *Modern Mechanix and Invention* (October): 70–78.

Russell, Jeffrey. 1997. *Inventing the flat earth: Columbus and modern historians*. Westport, CT: Praeger.

Schadewald, Bob. 1980. The Flat-out truth: Earth orbits? Moon landings? A fraud! says this prophet. *Science Digest* (July).

FLUDD, ROBERT

Elizabethan English philosopher, doctor, and astrologer who wrote a number of books on the relationship between man and the universe. Fludd was one of the last philosophers, prior to the scientific revolution, to try to link all of creation—both the macrocosm and microcosm—into a unified whole of knowledge leading to God. His best known work is *Utriusque Cosmi, Maioris scilicet et Minoris, metaphysica, physica, atque technica Historia* (1617–1621), commonly called the *History of Two Worlds*. As an astrologer and physician, Fludd worked to connect different parts of the human body to the astrological signs of the zodiac following sympathetic magic. He constructed a model of the universe combing the Ptolemaic geocentric scheme with Galenic medical concepts of the four humors. He worked out a rational, and illustrated, explanation of the creation of the cosmos and asked questions about what existed prior to the creation. He rejected Copernican Heliocentrism, arguing that if the earth orbited the sun, its surface would be swept clear by the winds generated by its movement, thus rendering the surface uninhabitable.

Robert Fludd, an influential alchemist, produced a comprehensive and intellectual story of the creation of the universe.

He also argued that cosmic power originated at the outer edges of the universe, rather than at the center. He pondered the structure of the human brain and bodily anatomy as well as music theory and meteorology. While ingenious and erudite, most of Fludd's impressive scholarly output was undermined by the discoveries and theories of the scientific revolution.

See also: Alchemy; Astrology; Metaphysics.

Further Reading

Godwin, Joscelyn. 1979. *Robert Fludd: Hermetic philosopher and surveyor of two worlds*. London: Thames and Hudson.

FORMAN, SIMON

An Elizabethan medical astrologer and social gadfly, Simon Forman (1552–1611) was also accused of being an alchemist and occultist. With no medical training, Forman set himself up as a physician in and around London, established a thriving practice, battled with the medical establishment, and generally lived an extraordinary life. He is best remembered for his voluminous manuscript writings and notes, almost none of which were published in his lifetime. He described his astrological medical work, musings on everyday life, his interest in the theater—including early mentions of Shakespeare—and other thoughts. Medical astrology used astrological charts and was rooted in the belief that heavenly bodies affect human health. Forman would cast a person's chart and horoscope and then provide remedies that often included the wearing of specially prepared talismans. Forman is considered one of the earliest known examples in the modern era of life writing, or autobiography.

See also: Astrology.

Further Reading

Kassell, Lauren. 2007. *Medicine and magic in Elizabethan London: Simon Forman: Astrologer, alchemist, and physician*. Oxford: Oxford University Press.

Traister, Barbara Howard. 2001. *The notorious astrological physician of London: Works and days of Simon Forman*. Chicago: University of Chicago Press.

Simon Forman was a flamboyant medical astrologer in Elizabethan London. He battled the medical establishment, who thought him a charlatan.

FORT, CHARLES HOY

American author who pioneered the study of anomalous or "damned" knowledge that did not fit standard patterns of scientific or theological belief. Fort (1874–1932) obsessively collected newspaper accounts from libraries in New York and London of odd occurrences. These included reports of falls of frogs, strange disappearances, unusual celestial phenomena, and other such material. Fort argued that such data was regularly rejected, or damned, by mainstream science because it did not fit preconceived notions or accepted explanations. He developed a theory that this unusual evidence suggested there were forces at work in the universe that scientific orthodoxy would not or could not accept. He argued that, if there was evidence of something happening, regardless of how strange or outside the norm it was, it had to be considered as possibly genuine. The reason the mainstream rejected this data was that they were wedded, he said, to their political positions rather than a belief in an objective study of nature. His proto-postmodern work was not well known during his lifetime, but found renewed interest in the later 20th century by readers ready to embrace a worldview at odds with dominant paradigms. Fort's work has been so associated with odd phenomena that strange coincidences, odd number relationships, spontaneous combustion—and many of the topics covered in this encyclopedia—are thought of collectively as Forteana.

See also: Forteana.

Further Reading

Fort, Charles, and Damon Knight. 1975. *The complete books of Charles Fort: The book of the damned / Lo! / Wild talents / New lands.* New York: Dover Publications.

Michell, John and Robert Rickard. 1979. *Phenomena: A book of wonders.* London: Thames and Hudson.

Steinmeyer, Jim. 2008. *Charles Fort: The man who invented the supernatural.* New York: Tarcher.

FORTEANA

Named for anomalist writer Charles Hoy Fort (1874–1932), Forteana constitutes a category of strange phenomena characterized by its seemingly inexplicable nature, numbers, locations, or frequency. Classic Fortean occurrences are falls of frogs or other objects from the heavens, unexpected coincidences, anomalous animals or fossils, and generally any phenomena that seems to defy or contradict accepted historical or scientific wisdom or the laws of nature.

See also: Anomalous Fossils; Fort, Charles Hoy.

Further Reading

Fort, Charles and Damon Knight. 1975. *The complete books of Charles Fort: The book of the damned / Lo! / Wild talents / New lands.* New York: Dover Publications.

Michell, John and Robert Rickard. 1979. *Phenomena: A book of wonders.* London: Thames and Hudson.

Steinmeyer, Jim. 2008. *Charles Fort: The man who invented the supernatural.* New York: Tarcher.

GAP THEORY

Popular creation science notion that between the biblical creation and the beginning of human history there was an indeterminate period of time, or gap, separating them. This allows for science-conscious Christians to reconcile the disparity of the Genesis account, which suggests a young earth, and the extensive geological evidence pointing to an old earth. Gap theory traces back to Scottish theologian Thomas Chalmers (1780–1847) who used it as a way of reconciling scripture with the knowledge geologists were then just beginning to produce about the nature and age of the earth. Following the Scopes Monkey Trial of 1925, in which Christian fundamentalism was thought to have been defeated, some anti-evolutionists moved away from a purely scriptural point of view and began building what they thought would be a scientific refutation of human evolution. Creationism is the belief that the earth and all life on it are the result of divine intervention. Creation science was a fundamentalist attempt to generate empirical evidence supporting a literal interpretation of the Genesis story of a six-day creation week occurring no more than 10,000 years ago, which could be proven by scientific methods. Not all creationists took the young earth position. Some allowed that the earth might be billions of years old and used gap theory as a way to support that idea. Strict literalists tend to reject the gap theory because it does not fit biblical accounts and thus contradicts God's revealed word. A related idea is day-age theory.

See also: Creation Science; Day-Age Theory.

Further Reading

Numbers, Ronald. 1992. *The creationists.* Berkeley: University of California Press.
Regal, Brian. 2005. *Human evolution: A guide to the debates.* Santa Barbara, CA: ABC-CLIO.

GAY REPAIR THERAPY

Program engaged in by Christian activists who believe both male and female homosexuality are psychological illnesses that can be cured. They also paradoxically believe homosexuality is a lifestyle choice. Proponents run psychotherapy-style encounter groups and one-on-one sessions and believe that, with proper counseling, religious instruction, and an acceptance of Jesus as one's personal savior, the sufferer will give up interest in same-sex encounters and become heterosexual. There is also a growing series of camps designed for the treatment of teenage homosexuals. The majority of Western medical personnel see gay repair therapy as not only useless—as homosexuality is not considered a disease in need of curing—but possibly dangerous physically, emotionally, and spiritually.

A central issue in discussions over homosexuality is whether it is a natural biological condition (homosexuals are born) or if it is simply a lifestyle choice (homosexuals are made). This is the nature versus nurture argument. Progay activists insist homosexuality is a normal biological condition into which people are born. The mainstream medical and scientific communities also take this position. Those who oppose homosexuality are mostly from religious backgrounds and do so from a theological standpoint, arguing that the Bible prohibits such behavior. Religiously based antigay advocates also resort to homosexuality as a disease approach.

In a recent vote concerning gay marriage rights in Spain, a professor of psychology, Aquilino Polaino, supporting the conservative antigay position, claimed that homosexuality was an abnormal pathology caused by bad parenting. Interestingly, the conservative "family values" party the professor was intending to support rejected the professor's position as archaic. This is worth noting, as in North America the forces of religious and political conservatism and "family values" do believe homosexuality is a disease. (Antigay factions cannot seem to agree if homosexuality is a lifestyle choice or a disease, but they do all agree it is wrong and abnormal.) The idea that being gay is a psychological abnormality goes back to antiquity. In the 20th century, homosexuals were routinely sent to hospitals and sanitariums to be "cured," based on the reasoning that they deviated from normal, therefore they were deviants. The American Psychiatric Association (APA) supported this position and considered homosexuality a pathology until 1973, when it reversed itself and removed homosexuality from its list of psychological disorders. More recently, the APA has stated that gay reparative therapy does not work and is dangerous.

Those religious advocates who believe homosexuality is a disease also argue that there are techniques for "curing" it. The main approach to curing homosexuality is to get the "sufferer" to get back to "normal" behavior and a closer relationship with God and Jesus. Therapy takes a psychological approach as well as abstinence and discussions of why being gay is "wrong" and against nature and God. There are a number of Christian organizations that specialize in gay repair therapy. The National Association for Research and Therapy of Homosexuality (NARTH) takes on a clinical persona stressing that what they do is good science and medicine and that theirs is not a subjective, antigay organization, but an objective medical one. Their carefully worded Web site says they "uphold the rights of individuals with unwanted homosexual attractions . . . and the right of professionals to offer that care." This way, the emphasis is on individuals wanting to be cured and is a case of individually freed patients requesting treatment from professionals who give it.

A major "healing ministry" that is a straightforward religious organization is Exodus International. Exodus International is an offshoot of self-appointed moralist Kevin Dobson's Focus on the Family organization, was run in the 1990s by ex-gay (a term coined in 1980) advocate John Paulk. An engaging speaker, Paulk toured the country at the behest of Focus on the Family telling about his personal conversion experience and how gays can be "cured" and made "normal" again at "Love Won Out" gay repair conferences and seminars. He published an autobiography *Not Afraid to Change* (2000). Shortly after the book's publication, however, Paulk was caught in a Washington, DC, gay bar and admitted he was never "cured" and was still gay. He claimed his work as an ex-gay spokesman was a sham of which he was ashamed.

If being gay is a natural state of affairs, like being blond or tall, it is beyond the ability of science to cure it, as it is not a disease. Gay repair therapy is an especially nefarious pseudoscience because it makes a normal person feel broken or damaged when he or she is not. This can cause great emotional and physical harm. Increasingly, ministries are operating ex-gay camps specially designed for teenagers sent there by their parents. A number of reports coming from these camps have led critics to argue they amount to organized child abuse. The mainstream medical and psychiatric community is in growing agreement of their opposition to gay repair therapy. Any attempt to "cure" a gay or lesbian person has as much chance of success as trying to "cure" a dog into being a cat.

Further Reading

Besen, Wayne. 2003. *Anything but straight: Unmasking the scandals and lies behind the ex-gay myth.* London: Routledge.

Paulk, John. 2000. *Not afraid to change: The remarkable story of how one man overcame homosexuality*. Spokane, WA: Hartline.

GEOMANCY

The "science of sand," geomancy is an occult technique in which marks on the ground, piles of dirt, and lines scribed in sand can be used for divination or fortune-telling. Geomancy was practiced as part of alchemy and astrology, but not by all practitioners. The word first appeared in English in the early 1300s. Geomancy works by reading the patterns in sand, dirt, or soil and interpreting them. It is similar to reading tossed bones, stones, cards, or other such objects as a form of fortune-telling.

Further Reading

O'Brien, Paul D. 2007. *Divination: Sacred tools for reading the mind of God*. Portland, OR: Visionary Networks Press.

GHOST HUNTERS

American reality television program on the cable SyFy Channel (formerly known as the Sci Fi Channel) about a group of young amateur paranormalists from Warwick, Rhode Island. The group, called TAPS (The Atlantic Paranormal Society) travels to different locations around the United States (and in a later incarnation, to Europe) investigating alleged hauntings. They go to a wide range of locations, from the well-known like the Lizzy Borden House and Winchester Mystery House, to local legends to private homes. The underlying narrative is that the TAPS team goes into an investigation trying to

debunk the allegations. In doing this, they claim, they can either show whether there is a genuine haunting or if the effects reported have a prosaic explanation. Their approach involves a number of methodologies they claim are scientific: the use of infrared night vision devices, audio and video recorders, and other electronic gear. During the course of any given investigation, normally one long night, they attempt to communicate with spirits alleged to be occupying the location. Sometimes they simply ask questions, other times they try to provoke a response by angrily demanding the spirits perform, the latter often involving liberal use of expletives that must be bleeped out for broadcast. The next day, team members laboriously go through all their electronic recordings for evidence of ghostly manifestations and present their findings to the owners of the location in a dramatic scene known as a "reveal."

Despite the apparent sincerity of the TAPS team and attempts to run their investigations scientifically, none of the group's members has any scientific or technical training. None have studied engineering, psychology, or the history of science. One positive aspect of the TAPS members' backgrounds is that team leaders and TAPS founders, Jason Hawes and Grant Wilson are plumbers by trade. Their expertise in that field allows them to quickly determine if a haunting effect is in reality a plumbing-caused effect (air thumping in water lines, drips, or slowly draining sewers), which it often is.

In their attempt to explain why they use certain techniques, TAPS members often resort to statements such as "it is believed that . . ." or "there is a theory that . . ." and "some paranormal researchers argue that . . ." without ever referencing their sources or how they have come to their conclusions. For example, a favorite technique of the TAPS team is to continually check the temperature of the air in the target location. They explain that "it is believed that" a ghostly spirit or similar entity's presence will cause the ambient temperature to drop. Suddenly rising or dropping temperatures are always a cause for great excitement among the team. Unfortunately, there is no attempt to discern why a spirit, if present, would cause a localized temperature drop. From a pseudoscience point of view, this attempt at scientific rationality is where the TAPS team often gets into trouble. While they take on the appearance of scientific objectivity, their methods have no basis in reality; there are no data to support the use of any of the techniques they employ. The team does not seem to keep track of their activities, such as charting times when temperature shifts coincided with apparitions or whether the voices they often record — through a technique they call "EVP work" — are disembodied voices or simply artifacts of the recording process. They often fail to show they even understand the rudiments of the equipment they employ.

Ultimately, all of these critiques could be made less glaring if TAPS ever found genuine evidence of a ghostly haunting. The voices they record could just as easily be put down to earthly causes, while the bumping and thumping sounds they often get excited about are clearly cases of squirrels in the attic. Their most famous piece of video footage was clearly someone under a sheet who had sneaked into the target site in order to confound the TAPS team. Many times during these investigations, someone on the team with professional science or engineering training could have helped alleviate confusion, streamlined the work, and maybe even produced genuine results. That would, however, defeat the purpose of TAPS. They are an amateur organization; professionals would have taken all the fun out of it.

Despite its scientific and logical shortcoming, *Ghost Hunters* is an entertaining program. The TAPS team members exude a kind of goofy sincerity and charm that is appealing. Besides the ghost hunting, the team engages in good reality television internal squabbles, arguments, and other interpersonal shenanigans. In an age where the larger run of people reject the work and knowledge of experts, a bunch a lovable, untrained, clueless enthusiasts running around hunting for ghosts holds greater authority than genuine scientists who investigate the natural world. The TAPS team are like latter-day Little Rascals who converted their barn clubhouse into a ghost-hunting society. TAPS is the club every kid wanted to form or join.

The number of TAPS-like groups has grown exponentially since the show first aired in 2004, and the show's success helped spawn other like-minded imitators: the British based *Ghostly Encounters*, *Dead Famous*, which only looks for the ghosts of celebrities, and *Paranormal State*, a more pompous and self-consciously serious copy of *Ghost Hunters*.

See also: Ghost Hunting.

GHOST HUNTING

The act of searching for evidence of ghosts, poltergeists, and other spirit entities. Some prefer the term investigator rather than hunter. There are two types of ghost hunters, those looking for evidence to support the existence of spirits and those seeking to debunk the idea; on occasion it is a combination: True believers wish to eliminate false reports and hoaxes so that only genuine examples are known, while debunkers attempt to show there is nothing there at all. Investigators use an array of technical devices like temperature gauges, ion meters, electromagnetic sensitivity meters, infrared cameras, motion detectors, and recording devices of all types.

Eighteenth-century ghost hunters dealt mostly with literary sources. It was in the 19th century that active pursuit of spirit entities began. Investigators used photography to capture ghosts' images in spirit photography. They also studied accidental captures of ghosts on film and tried to account for them. One of the earliest scientific group efforts to investigate ghosts was the French-based *Institute Métaphysique International* of 1919. Begun by academics including Charles Richet, the man who coined the term "ectoplasm" and Gustave Geley (1865–1924), the author of *L'etre Subconscient* (1899), who argued for soul energy and believed psychic ability genuine, the *Institute Métaphysique* investigated a range of paranormal activities including the "Kluski Hands." Polish-medium Franek Kluski (1874–1944) claimed he could summon spirits whom he could convince to leave traces of themselves, then he would produce ectoplasm. A test was set up with bowls of liquid wax. Kluski was to ask the spirits to plunge their hands and feet into the warm wax. A void was created and then plaster poured in making a cast. This test produced a collection of plaster hands and feet supposedly made by actual ghosts. Artifacts such as these are termed permanent paranormal objects (PPOs).

Various current proghost societies are made up mostly of amateur investigators with only minimal scientific training. Organizations like the Committee for the Scientific Investigation of Claims of the Paranormal (CSICOP) are made up of trained personnel, but are normally skeptical debunkers rather than believers. The model of the modern amateur ghost hunter and aspiring scientific investigator was the Englishman Harry Price (1881–1948). Price began his career as a magician and spiritualist, and later took to studying how magic was performed, before moving onto investigating claims of paranormal activity. He was instrumental in exposing a popular, but oafish spirit-photographer William "Billy" Hope in 1922. Price managed this by having the arrogant Hope perform his work under a series of controlled tests Price had worked out. He performed this study after joining the *British Society of Psychical Research,* an organization inspired by the French *Institute Métaphysique.* Price was more investigator than debunker, as he did believe psychic powers existed and wanted to prove it. He had no science training, but enjoyed taking on the persona of the serious research scientist. He employed the now-standard ghost-hunting technique of checking ambient air temperature in a haunted place, but unlike most later investigators kept careful records of his investigations and documented séances. He eventually became the head of the *National Laboratory of Psychical Research* in 1925. Unlike his contemporaries, Price was willing to embrace the media and used it to his advantage to further the cause of psychic research and awareness.

Price's most famous project was his investigation of the infamous Borley Rectory of Essex, England. A reputedly haunted parson's house said to be the site of considerable paranormal activity, Borley had earned the reputation of "most haunted house in England." In 1930, a new rector named Lionel Foyster and his wife Marianne arrived. This signaled the start of intense spirit activity. Poltergeists (violent spirits) began throwing things around, including bricks and dishes. It was said a monk and a nun had died in the rectory trying to elope and it was their restless spirits who were performing the haunting. Mr. Foyster reported finding writing on the wall from the nun. After the Foysters moved out in 1935, Price took over and began investigating. He solicited almost 40 volunteers to set up surveillance and crafted a handbook for them as a guide for how to conduct a paranormal investigation. The investigation produced evidence of what Price felt was the ghost of a murdered woman on the grounds. Price produced two books *The Most Haunted House in England* (1940) and *The End of Borley Rectory* (1946). Later, there were allegations of fraud on Price's part related to a series of photos taken while the rectory was being demolished in which bricks seem to fly. Price has been alternately been called the "father of modern ghost hunting" and a con man. He has certainly had many followers in the United Kingdom and the United States.

Late-20th-century ghost hunters include the Committee for Skeptical Inquiry (CSI) and CSICOP. Both work to debunk paranormal claims and look for frauds and scams behind such claims. CSICOP has been accused of being merely an antiparanormal propaganda machine. A number of universities around the world have parapsychology research laboratories and a number now offer advanced degrees in the field. They test claims of paranormal activities made by individuals and groups in controlled laboratory conditions. Rather than go to sites of para-normal activities like hauntings, these researchers tend to be less dramatic and more scholarly. They are interested in how the human body, and especially the mind, might produce such effects and how biological function might be misinterpreted as paranormal. Such institutes can be found at The Rhine Research Center at Duke University, and the Koestler Parapsychology Unit (KPU) at the University of Edinburgh, Scotland.

The central conundrum of ghost hunting is that there is no one accepted definition of what a ghost is. There are many long-standing traditions that ghosts are spirit entities or the astral form of a deceased person, but these definitions are vague at best and do not address the underlying issue of life after death, the presence of a soul, or a life form separate from the body. There is nothing more than anecdotal evidence that the soul—whatever it is— continues to exist after physical death. If it does, we have no coherent hypothesis about why it would "haunt" an area. What is the process of ghost creation? Does it happen at the moment of death, or is it a longer process? Until points such as these are addressed, ghost hunting will remain an unorganized exercise in futility.

See also: Ectoplasm; *Ghost Hunters*.

Further Reading

Geley, Gustave. 2003. *Clairvoyance and materialization: A record of experiments*. Whitefish, MT: Kessinger Publishing.

Hansen, George P. 1992. CSICOP and the skeptics: An overview. *The Journal of the American Society for Psychical Research* 86 (1): 19–63.

Morris, Richard. 2007. Harry Price investigates. *Fortean Times* 229:28–34.

Price, Harry. 1942. *Search for truth: My life for psychical research*. London: Collins.

HERMES TRISMEGISTUS

Name given to the legendary father of alchemy and Western occult mysticism, it is a combination of the Greek god Hermes and the Egyptian Toth. The name is often translated as "thrice great Hermes," referring to the fact that there are several Greco-Roman deities named Hermes or Mercury. The mythical figure is sometimes cast as the child of gods or, as in the Kabbalist tradition, as a contemporary of Moses. In this case, Hermes is seen as an especially gifted magus handing down wisdom and knowledge to adepts. Renaissance physicians and alchemists thought Hermes an all-knowing ancient Egyptian who was the author of the heavily influential *Corpus Hermeticum.* Historians today argue Hermes was not a single individual and that the *Corpus* was not written in the mists of the ancient past, but more likely during the second and third centuries AD. The *Corpus Hermeticum* combined extant Greek, Persian, Jewish, and Christian mysticism. It was initially separate from operational alchemy and astrology, but eventually combined with them to add a metaphysical dimension to the more practical aspects of chemistry and stargazing. The ideas and teachings of Hermes Trismegistus are known collectively as Hermetica and the study and practice of those ideas as Hermeticism.

Renaissance alchemists were especially enamored of Hermes and looked to him as the model alchemist and his writings as a primary authority. These works included texts on alchemy, astrology, magic, and medicine as well as philosophical discussions of the creation of the universe and the nature of man. They were codified into a collection of 42 primary texts. Francis Bacon separated them into two categories: theoretical and practical alchemy; the first included philosophical musings, while the second discussed practical recipes, potions, and procedures.

The *Corpus Hermeticum* first came to wide European notice after it was translated into Latin by the Italian philosopher and astrologer Marsilio Ficino (1433–1499). A scholar of wide-ranging interests, Ficino also made the first Latin translation of Plato and accepted the largely Greek origin of the Hermetic text. He wrote on the nature and immortality of the soul and, as a Neo-Platonist, believed all living things, including the earth and heavens, had souls that were intertwined and connected. His translation was widely reproduced and influential. Enthusiasts came to believe the book was written by Hermes in ancient times and thus represented a long-lost mystical tradition, but in 1614, French philologist Isaac Casaubon (1559–1614) showed the *Corpus* could not be as ancient as was popularly believed. A careful study of the grammar used suggested to him that at least some of the texts—the more philosophically minded ones—were a bit too modern. He suggested they were written in the early centuries of the modern era, and probably by Greek-speaking authors living in the Roman Empire. (There are anti-Roman sentiments sprinkled throughout the texts.) Other, later texts were written in north African and Middle Eastern languages, including Coptic and Arabic.

While no direct references are made to Jewish or Christian sources in the *Corpus,* there are sections that are clearly influenced by such works. The first chapter or book of the *Corpus Hermeticum* is known as the *Pymander* (it is spelt differently in a number of versions). The character of Pymander is known as "the Sheppard of men," and may very well be God. Most of the *Corpus Hermeticum* is written in the form of a dialogue between a master and an aspiring adept. The text lays out the origin of the material world as created by a nonmaterial and unknowable god. It also discusses man's place in the universe. The text argues that man's fall began with the divinely created material universe. Men must somehow escape the material world in order to get back in touch with the realm of the spirit. More importantly, it discusses the paradox of an all-powerful god who creates a material universe it does not like and which it then tries to eliminate.

The most popular and influential book of the *Corpus Hermeticum* is the *Emerald Tablet.* This text is said to be a secret recipe for a primordial substance Renaissance enthusiasts believed was key to making the Philosopher's Stone. It contains the idea that would be made popular with the phrase, "As above, so below." Organized Christianity took an indulgent view of the *Corpus* because parts of it, like the *Pymander*, could be seen as presaging the Bible. If accepted as a book written in the time of Moses, the *Corpus* was not inconsistent with Christian ideals. Even the *Emerald Tablet* could be argued as an allegory about finding God in one's heart. The rage for alchemy the *Corpus* engendered thus brought on little scrutiny for a time from Church authorities. The mistaken belief that the *Corpus* dated back to pre-Greco-Roman times, even to Moses himself, contributed to the acceptance of the work. What enthusiasts were doing, however, was imbibing ideas from the very age they sought to repute. The Renaissance magi thought they had turned away from the rationalism of the classical age, Aristotle, and Plato, to embrace a Christian friendly mysticism, but they had unwittingly taken on the classical concept of reason. It can be argued that this misunderstanding helped lead to the very kind of embrace of rationalism and general rejection of the supernat-

ural and mystical that so embodied the Enlightenment.

Some modern-day occultists continue to insist that there was a genuine person named Hermes Trismegistus. They insist that his texts are of Egyptian origin and are kept in a hidden library accessible only to a few exceptional adepts. His memory has been linked to Atlantis, lost civilizations, hidden history, and a host of esoteric pursuits. In the late 19th and early 20th centuries, an occult revival flowered that was heavily invested in Hermeticism. A number of occult groups, largely of British origin, appeared. The Rosicrucians, the Golden Dawn, the Theosophical Society, and others took the *Corpus Hermeticum* at face value, including taking on the trappings of Egyptian-influenced art, music, and costume. Self-styled adepts and magi like William Robert Woodman, S. L. MacGregor Mathers, Madame Blavatsky, and, most notoriously, Aliester Crowley, brought Hermetic ideas to a wider audience and began a craze for esotericism that has yet to abate.

See also: Alchemy; Astrology.

Further Reading

Segal, Robert A. 1986. *The poimandres as myth: Scholarly theory and gnostic meaning.* New York: Walter de Gruyter.

Thorndike, Lynn. 1947. *A history of magic and experimental science.* New York: Columbia University Press.

Yates, Francis. 1964. *Giordano Bruno and the hermetic tradition.* Chicago: University of Chicago Press.

HEUVELMANS, BERNARD

Bernard Heuvelmans (1916–2001), along with Ivan Sanderson, is considered the father of cryptozoology, the search for hidden animals. Born in France but raised in

The French-born zoologist Bernard Heuvelmans is the cofounder of cryptozoology.

Belgium, Heuvelmans earned a doctorate in zoology in 1939 from the Université Libre de Bruxelles. Though his doctoral thesis was on aardvark teeth, the mythological aspects of zoological history fascinated him. He began collecting reports and other materials about such animals as griffins, mermaids, and others that had achieved legendary status as something other than what they were. Unable to find work as a zoologist in war-torn Europe, Heuvelmans worked variously as a jazz musician, a stand-up comedian, and a writer. In 1948, he read the *Saturday Evening Post* article by Ivan Sanderson "There Could Be Dinosaurs," in which the author suggested local legends supported the idea that extinct animals had survived to modern times. Heuvelmans then began to consider publishing his own work and going back to the career of his training. The result of this early period of wider research and inspiration was *Sur la Piste des Bêtes Ignorées* (1955) (released in an English translation in 1958 as *On the Track of Unknown Animals*).

The book established Heuvelmans as an authority on what came to be called cryptozoology and was the standard text by which all others were judged.

See also: Cryptozoology.

Further Reading

Heuvelmans, Bernard. 1955. *Sur la piste des bêtes ignorées*. Paris: Librarie Plon.
Sanderson, Ivan. 1948. There could be dinosaurs. *Saturday Evening Post*.

HIDDEN HISTORY

The idea that there is a vast unknown storehouse of knowledge that has been forgotten or suppressed. This knowledge is usually said to be of a glorious past, a social height that humankind achieved but for various reasons has lost. Hidden history is allied with romantic notions of the past as a fantasy more attractive than the reality of the present; it is the concept of the golden age. Golden age theories almost always include not only a physical decline, but also a loss of precious knowledge that only the purity of a return to past conditions would allow. Talk of lost wisdom and hidden histories is juxtaposed with contempt for the modern industrial world with its loss of spirituality and prevalence of corrupt Western decadence. Hidden knowledge remains hidden, not only because of its great age but also because of modern forces that work actively to keep it hidden. These nefarious forces want to keep this information out of the hands of those who would bring it to the masses or use it for their own purposes. It is also hidden because its acknowledgment would undermine accepted modes of thought. Less sinister is lost wisdom that was known to the ancients, but over the eons has simply been forgotten. The knowledge is made visible today in ancient buildings like the pyramids, artworks on cave walls and temple ruins, and in bits and pieces of knowledge retained to the present by still primitive tribal peoples who possess knowledge of heavenly bodies or medical cures only recently "discovered" by modern science.

Part of the hidden knowledge concept is the collecting of empirical evidence that supports the "lost" wisdom and contradicts mainstream thinking. It is commonly held that the mainstream is aware of the existence of contradictory data but ignores it or keeps it suppressed. Possibly the greatest collector of such data was Charles Fort (1874–1932). A sort of thinking man's Robert Ripley, Fort spent most of his life gathering and pondering what he called "damned" knowledge. He looked for evidence of experiences that did not fit contemporary scientific and intellectual paradigms and as such had been discarded, or damned, by science. He ransacked the world's newspapers at the New York Public Library and the British Museum for accounts of strange phenomena, artifact discoveries, and other anomalous flotsam and jetsam of the human experience. Taken in the aggregate, Fort believed such reports constituted an alternative body of data not easily waved off by mainstream thinking. For him, modes of analysis and theoretical frameworks came and went, but facts always remained to challenge explanation. Classic lost wisdom stories are evident in myths of the lost city of Atlantis.

The most successful modern purveyor of the hidden history concept is British journalist turned anomalous history investigator Graham Hancock. His works include *The Sign and the Seal* (1992) and *Fingerprint of the Gods* (1995), in which he argues that a lost, great civilization thrived prior to the beginning of the current written record and that it is the basis of all later civilizations. As with all hidden history aficionados, Hancock holds to the line that experts and other academics (he has no training in science or archaeology) are engaging in a major cover-up of humanity's golden past. He sees himself as part of a noble coterie of amateur investigators who are all that stands between humankind and the forces of darkness. He takes a catastrophist view of human history, arguing that the human experience has been altered, destroyed, and begun by various violent and sudden geologic changes in the form of earthquakes, floods, and pole shifts. The most recent such pole shift occurred around 10,450 BC and was responsible for the destruction of most of the golden civilization and the scattering of the survivors, who became the gods of ancient mythology and religion.

See also: Atlantis; Fort, Charles Hoy; Pseudohistory.

HOLLOW EARTH

The belief that the earth is a hollow sphere with life of some sort living inside it. A popular contrivance in literature, ideas about the earth being hollow were either aided by or aided from early religious concepts of an underworld. Christian theology, for example, indirectly accepts a hollow earth with notions of Hell. This is the abode where the devil and his minions dwell along with the souls of the dead who did not live in a manner suitable to gain them entrance to heaven. Not quite the same, the Greek concept of Hades is an underworld. Prescientific revolution concepts of a hollow earth are legendary, mythical, and religion-oriented rather than fact-based. There have, however, been several serious postscientific revolution arguments made for a hollow earth.

The most influential nonfiction theory of a hollow earth was put forward by war hero John Cleve Symmes II (1780–1829). The Symmes family had a long history in early American politics—John Cleve Symmes I was a Continental Congress delegate and helped open the Ohio territory to settlement. John Cleves Symmes often referred to himself as "Junior," though he was the nephew, not son, of the Ohio pioneer. As a young man, Symmes joined the army and rose through the ranks to captain and fought with distinction in the War of 1812 along the Canadian border. After his service, he moved to Kentucky and began to construct his hollow earth theory. He got it into his head that a sphere filled with "loose matter" would react a certain way. He said gravity and centrifugal force would force the interior fill of a sphere to form into a series of concentric shells inside an outer shell. If this was the case, he reasoned, the earth should conform to the same phenomenon. He also argued that it would be ridiculous for God to have created a solid earth, thus wasting all that interior space. Based upon this reasoning, he put forward the notion that the earth was hollow, that the interior contained possibly four to six concentric spheres, and that each sphere, including the outermost one, had circular openings at their poles to allow in light and air. He first brought his idea to public attention when he ran a short newspaper item in *Nile's Weekly Register* in June 1818 (he also claimed to have sent out copies to every institution of higher learning in America). He declared his belief in a hollow earth and said that he was preparing an expedition to explore the opening in the North Pole. Like Jason of Greek mythology, Symmes called for a group of "champions" to accompany him. He also began to query state and federal legislators about sponsoring the expedition. Like so many promoters of fringe science ideas, Symmes felt put upon by skeptics, so in February 1819 in a fit of irritation, he issued a challenge to "any opposers of my doctrine, to shew [*sic*] as sound reasons why my theory is not correct, as I can shew why it is." He did not offer a reward. Symmes gained converts, but little financial aid. There is some doubt as to whether he gained any support at all in Congress, as legend claims. In 1824, an Ohio newspaper ran a series of letters debating the idea. Symmes and his family had such a good standing in the community that he was not ridiculed locally, though that was not the case nationally, where the opening in the earth's surface was derogatorily referred to as the "Symmes Hole." One of Symmes's most energetic acolytes was James McBride (1788–1859), an amateur archaeologist and antiquarian who also investigated the Indian mounds of the Ohio Valley. McBride published *The Theory of Concentric Spheres* (1826) explaining Symmes's idea in detail. Symmes himself published only a handful of short newspaper items expounding his theory. The one possible exception may be *Symzonia: A Voyage of Discovery* (1820), in which Symmes used the pseudonym Captain Adam Seaborn. Detractors claimed Symmes had written the novel. Editor J. O. Bailey, who wrote the introduction to the 1965 reprint of the book, felt sure Symmes had written it. However, there is debate about whether Symmes did indeed write this book. (It is also a matter of question as to whether James McBride wrote the *Theory of Concentric Spheres*.) Professor William Marion Miller thought it unlikely that a man of Symmes's character and sense of honor—he was a duelist—would have ridiculed himself and his work just to propagate it. William Stanton, literary critic and author of *The Great United States Exploring Expedition* (1975) believed with certainty that Symmes did not write *Symzonia.* He took his position by comparing the grammar and stylistic content of the text to that of Symmes's known writings and found them incompatible. It has been argued that Seaborn may have been the pseudonym of author Nathaniel Ames (1764–1835), who wrote books on sea voyages and was himself a sailor who had traveled and read widely. Passages in Ames's *A Mariner's Sketches* (1830) are reminiscent of *Symzonia* and the style of writing is the same as Ames. The book did include an illustration of the hollow earth that conforms to Symmes's description, so regardless of who wrote *Symzonia,* it is based upon Symmes's work. When Symmes passed away, his son, Americus Symmes, put up a memorial stone on his father's grave that is topped with a stone sculpture of a hollow earth. Americus put out his own version of his father's work and tried to keep it going. Desperate to save the reputation of his beloved father, Americus claimed that a number of the arctic expeditions that had gained fame at the time had, in fact, unwittingly traveled into the interior, which was news to those expeditions. *Symzonia* and Symmes's work did influence later writers, most notably Edgar Allan Poe in *The Narrative of Arthur Gordon Pym of Nantucket* (1838) and Jules Verne, with *A Journey to the Center of the Earth* (1864).

A variation of the Symmes idea was the notion put forth by failed medical doc-

This is the Symmes model of the hollow earth. The large openings at the poles were derisively called "Symmes Holes."

tor Cyrus Teed (1839–1908). He argued not only that the earth was hollow, but that human civilization and the earth's "surface," as it was commonly thought of, was on the inside of the globe looking in, rather than on the outside looking out. Teed developed a complex view of life and a quirky, religious cosmology bordering on the delusional. He attempted to combine alchemy, electromagnetic theory, reincarnation, and ideas about the nature of God. After accidentally electrocuting himself, Teed claimed a beautiful female angel came to him and informed him that he had been chosen to be a prophet; in fact, he believed he was the second coming of Christ. He adopted the more theologically appropriate name of Koresh and began to attract followers, calling his new religion Koreshanity. The central tenet of Koreshanity was that the earth was hollow, the outside universe nonexistent, and the sun, being at the center of the interior sphere, was an electrically powered device. The globe, he said, was 8,000 miles across and 25,000 miles in circumference. He laid all this out in his book *The Cellular Cosmology: The*

Earth a Concave Sphere (1905). Teed's ideas spread and he gained followers in a number of American cities including Chicago, Portland, and Denver. They eventually established a utopian community in Florida, called Estero, which included a bakery, cement plant, and power generator—Teed was especially interested in the transmission of electricity. Ever a gregarious communist, Teed's electrical equipment supplied power to the local community. In 1904, Teed got into an altercation with a local sheriff who beat him badly. He never really recovered from his injuries and died in 1908. His followers expected his imminent resurrection, but after some time, officials forced them to bury him. Following Teed's death, Estero began to die as well. The final few Koreshanity followers gave the town to local authorities in 1961, and it was eventually opened as a museum and state park.

One of the most recent expounders of hollow earth theory was Raymond Bernard. In 1969, he released *The Hollow Earth*. In it, he borrowed from almost every previous theorist, though mostly through unaccredited means. He took the opening at the poles concept of Symmes, the interior sun of Teed, and also threw in UFOs. Bernard then added a racist edge, saying that the "races" of the interior had contributed to the ancestry of the Chinese, Egyptian, and Eskimo peoples, whom he called the "brown races." He used everything but the kitchen sink to construct his hollow earth–centered world: conspiracy theory, hidden history, evil governments, doltish masses, and even Wilhelm Reich and Orgone. A German esotericist, Bernard's real name was Walter Siegmeister. He was a Rosicrucian and alternative medicine advocate who changed his name because of a run in with the U.S. Food and Drug Administration. He also had

ideas for a eugenics-based way to create a master race. More recent theorists have also argued the earth's interior was the abode of UFOs and of escaped Nazis, and of escaped Nazis riding UFOs.

See also: Eugenics; Flat Earth; Hidden History.

Further Reading

Lang, Hans-Joachim and Benjamin Lease. 1975. The authorship of *Symzonia:* The case for Nathaniel Ames. *New England Quarterly* 48 (2): 241–52.

Standish, David. 2006. *Hollow earth.* Cambridge, MA: Da Capo Press.

Symmes, Americus, ed. 1878. *The Symmes' theory of concentric spheres: Demonstrating that the earth is hollow, habitable within, and widely open about the poles.* Louisville, KY: Bradley and Gilbert.

Symmes, John Cleve. 1819. Arctic memoir. *National Intelligencer* (February).

HOMEOPATHY

Homeopathy is an alternative medicine that employs the law of similars ("like cures") in its remedies. Instead of confronting an illness using symptom suppressing drugs, homeopathy treats ailments by introducing the same allergen which is causing the problem. The theory behind this practice states that a small dose of a particular substance will cure the same symptoms that it produces in large doses. If there is a substance that produces symptoms of disease in a healthy person, the same substance in minute amounts can cure those symptoms in a sick person. It is claimed that these remedies, composed of highly diluted plant, animal, and mineral substances, stimulate the body's defenses so that it may heal and protect itself. This method is used to treat many conditions and it has been in continuous use in Europe for over 200 years. It was in the vein of the Hippocratic "law of similars" teachings that German physician Samuel Hahnemann (1755–1843) first developed the technique. He conducted his initial experiment on himself and was able to show that Peruvian bark, which contains quinine, could not only be used to treat malaria, but also caused malaria-like symptoms in a healthy person.

Homeopathy is considered a holistic form of medicine that helps the body heal itself. It can be used as a treatment for chronic and acute ailments, and to prevent illness. Instead of suppressing symptoms of disease like conventional allopathic medicine, it introduces minute quantities of allergens. Adherents believe this method encourages the body's immune response to heal itself. There are between 2,000 and 2,500 homeopathic remedies. Each preparation contains minute amounts of an individual substance which is diluted with milk, sugar, or alcohol and then shaken. This process is call *potentization* and is repeated up to 200 times or more. The further the substance is diluted, the more potent it is thought to be. A remedy may also be administered in pill form, by being placed directly under the tongue, where it can dissolve and easily enter the bloodstream.

When prescribing a remedy, the homeopath considers the patient's personal life, habits, emotions, diet, exercise, sleep patterns, complexion, appetite, moods, libido, posture, environment, and the weather, along with the symptoms they exhibit. While the symptoms may be the same in many people, the treatment for each person will vary. Homeopathic medicine is considered safe for everyone, including babies, but is only to be taken as it is needed. It is used to treat almost every physical condition. It is administered in emergencies to help with shock or to encourage the healing of injuries, and is also commonly called upon to treat emotional conditions like panic, anxiety, and fear. A typical treatment requires a consultation and a return visit once or twice a month to assess and adjust the prescription. Chronic conditions require extended treatment, while acute disorders may respond after just one visit.

The American Institute of Homeopathy, the United States' oldest medical society, played a significant role in the country at the turn of 20th century, when 15 percent of practicing physicians were homeopaths. At that time, the nation housed 22 homeopathic medical schools, including ones at Boston University and New York Medical College, 100 practicing hospitals, and over 1,000 homeopathic pharmacies. However, the paramount advancement of conventional pharmaceuticals eventually eclipsed homeopathic medicine. The formation of the American Medical Association, and later the Food and Drug Administration, allowed allopathic medicine to advance and preside over homeopathy using political and legislative power.

In recent years, homeopathic medicine has experienced a resurgence of popularity. It is widely used and widely available in Western countries. Studies in the well-respected *Journal of Pediatrics* and *Lancet* have proven that it is effective, and a review of multiple studies compiled by the *British Medical Journal* also concluded that the use of homeopathy had positive results. However, it has yet to be accepted as a standard, mainstream form of treatment because there has been no scientific way of explaining how or why it works. The mystery of homeopathy's function has to do with levels of dilution. The more diluted the tincture, the more powerful the results, yet in these concoctions, the infinitesimally small amounts of a substance can be diluted to the point where no original molecules remain. This leaves questions as to the true chemistry of the substance and how it produces the claimed healing results.

Those familiar with holistic and Eastern medicine profess that the healing effects of homeopathic medicine are caused by a remedy's subtle energy, which influences a person's vital force. Dana Ullman asserts that the substance leaves a "holographic imprint" of energy in the dilution which, in turn, acts upon a person's energetic qualities. She claims that this corrects a person's energy field, allowing the body to function with a higher degree of efficiency and to eliminate the manifestation of symptoms. These hypothetical qualities of electromagnetism have not been proven or understood. Similar concepts, such as the *chi* in Chinese medicine or the *prana* in Ayurveda, are widely accepted as the key to healing in the East. Western science has no parallel concept and does not accept it as a legitimate form of treatment.

See also: Alternative Medicine.

Further Reading

Horrocks, Thomas A. 2008. *Popular print and popular medicine: Almanacs and health advice in early America.* Cambridge: University of Massachusetts Press.

MacEoin, Beth. 2007. *Homeopathy: The practical guide for the 21st century.* London: Kyle Cathie.

Starr, Paul. 1984. *The social transformation of American medicine.* New York: Basic Books.

HORSE-RIPPING

Horse-ripping refers to assaults on horses and cattle in the form of mutilations and killing. For the purposes of the study of pseudoscience, there are two forms of horse-ripping. The more common is the assault upon horses for criminal and psychological reasons. Some individuals feel compelled to injure horses. As this is a pathological malady, it will not be considered here. The other form of horse-ripping, also called cattle mutilation because several forms of livestock are involved, began in the western United States. Large animals—horses, cattle, and sheep—are, on occasion, found dead with inexplicable wounds. Large chunks of flesh and organs are removed, often with seemingly surgical precision. Genitalia are not so much mutilated as removed and taken away. There is usually little or no blood at the scene, and perpetrators are never caught in the act. Researchers have claimed that the wounds are not caused by normal predation or scavenging. Because of the general strangeness of the condition of the bodies and the remote locations of the events, some have attributed horse-ripping to Satanic cults, cryptid predators, UFOs, and other sources.

The American ranching community first noticed horse-ripping in the 1960s. The first widely publicized case was in 1967 and concerned a Colorado ranch horse named Lady. The media wrongly attributed the event to another horse on the ranch, named Snippy, and his name was commonly attached to the story. In the literature, as here, Lady was often referred to as Snippy. The horse was found with the flesh and muscle removed from its head and neck. The horse's owners had never seen such wounds. An examination of the area around where Snippy was found showed evidence of burn spots, and a test by a local forest ranger found high levels of radiation. The ranger also said he had never seen such a thing in his experience. Others claimed to have seen UFOs in the area and strange lights in the sky. Investigators from the University of Colorado, who were attached to the UFO-focused Condon Committee, arrived on the scene and, like the others, claimed no earthy agent could have made the wounds. After the Snippy story went out over the news wires, many similar cases began to be reported across the United States and eventually around the world.

In addition to the elements reported in the Snippy case, there are a number of details associated with horse/cattle ripping: Reproductive organs are removed, the ripping has a clinical look and seems to be done with a scalpel or even medical laser, and the work centers on the head and genital areas. Common signs of predation and scavenging are absent. The attacks occur only in rural and isolated areas. When victims have been studied by state police forensics agents, university veterinarians, and other qualified persons, strange details are found, but nothing conclusive as to whom, how, or why is revealed. Unlike many anomalous phenomena reported by civilians, the horse and cattle ripping cases of the 1970s were investigated by the authorities. This may have stemmed from the fact that the cattle industry had many friends in both federal and local governments. While these investigations found peculiarities in the cases, they did not find anything to suggest the reasons for the deaths were other than natural. Fortean researchers have focused upon three basic categories of explanations: Satanic cults, secret government experiments, and UFOs. All three groups have their supporters and all have questions as to their veracity and plausibility. All three are held up as panaceas.

One of the more vocal proponents of the mystery of horse-ripping is investigative reporter and award-winning documentarian Linda Howe. She has investigated radioactive water supplies, the environment, child hunger, crop circles, and UFOs as well as cattle mutilations. She is convinced something strange is going on and that the wounds inflicted in these cases defy natural causes or even explanation. She also believes UFOs are involved. There are ranchers and wildlife agents, however, who argue the wounds inflicted on the horses and cattle are the results of predation and scavenging and suggest nothing paranormal or supernatural is afoot. Recent accusations claim the mutilations are being done by nefarious government agents working on mad cow disease and anthrax chemical warfare experiments and how to make it possible to get these toxins to jump across species and into the human population.

See also: Alien Abduction; Unidentified Flying Objects (UFOs).

Further Reading

Donovan, Roberta. 1976. *Mystery stalks the prairie.* n.p.: T.H.A.R. Institute.

Howarth, Leslie. 2001. *If In doubt, blame the aliens!: A new scientific analysis of UFO sightings, alleged alien abductions, animal mutilations and crop circles.* Bloomington, IN: AuthorHouse.

Howe, Linda. 1989. *Alien harvest: Further evidence linking animal mutilations and human abductions to alien life forms.* Albuquerque, NM: Linda Moulton Howe Productions.

HUMAN-ALIEN HYBRIDS

Supposed offspring of forced unions between humans (usually women) and extraterrestrial beings. Modern claims of

The gray alien is the most commonly reported extraterrestrial form.

aliens using human genetic material began with the Betty and Barney Hill abduction case of 1961. Since then, it has become a staple element in alien abduction stories. Female abductees now report having reproductive tests done, biological material removed, and alien materials injected. The abductee is often shown a human/alien hybrid child and told it is her offspring.

The great problem with the human-alien hybrid scenario is that, if sentient beings from other worlds are visiting the earth, the chances of them being the same species as humans is remote. Earth biology shows that, with very few exceptions, cross-species breeding is impossible. Those few exceptions (a horse and donkey mating being the most common) can only produce sterile offspring (in the form of a mule, to continue with the above-referenced example). The closest genetic relatives to humans are the higher primates, such as chimpanzees and bonobo apes, and they are incapable of cross-breeding with humans. As a result, any union between a human and an alien would have as much chance of producing offspring as the union between a human and a grasshopper.

Alleged proof of human-alien offspring is the "Star Child" brought to light in 1999. Supposedly found in a cave in the Mexican state of Chihuahua in 1930, the Star Child—whose skeleton consists of cranial fragments—came into the possession of anomalist writer Lloyd Pye, who began to promote it as the remains of a human-alien union. Genetic tests done in California, British Columbia, Canada, and London have shown the DNA to be consistent with known Native American peoples. Critics argue the strangely inflated skull is of a four- or five-year-old child who lived about 900 years ago and who suffered from progeria, the so-called aging disease, hydrocephaly, some other genetic disorder, or that it was possibly a victim of cranial binding, which is known to have been practiced among Native Mexican populations.

See also: Alien Abduction; Unidentified Flying Objects (UFOs).

Further Reading

De Lafayette, Maximillien. 2008. *Hybrid humans and abductions: Aliens-government experiments*. Scotts Valley, CA: CreateSpace.

Mack, John. 1995. *Abduction: Human encounters with aliens*. New York: Ballantine Books.

Pye, Lloyd. 2007. *The starchild skull—genetic enigma or human-alien hybrid?* n.p.: Bell Lap Books.

HYDRICK, JAMES ALAN

Hydrick was a self-styled psychic who claimed the power of telekinesis—the ability to move objects with only the power of his mind. In the late 1970s and early 1980s, Hydrick developed a following in and around Salt Lake City, Utah, by claiming he had perfected his powers through martial arts training with Chinese Kung Fu masters. He came to worldwide fame late in 1980 when he demonstrated his powers on television. The next year, he was shown to be a fraud by James Randi—again on television—and fell into disrepute. Shortly after, he confessed to performing the tricks through controlled breathing and dropped out of sight.

James Hydrick first came to the public eye when he appeared on the American television program, *That's Incredible!* He performed feats of telekinesis by moving objects, including the pages of a phone book. Tabloid news accounts hailed Hydrick as the best psychic in the world and he was feted as a natural healer who could cure headaches and read people's minds with his own. He was then invited to perform on *What's My Line?* along with paranormal investigator James Randi. Hydrick was first allowed to perform with no restrictions and executed his pencil and phone book page moving to the delight of the crowd. Then, James Randi sprinkled Styrofoam particles in front of the book and asked Hydrick to repeat the feat. Randi was convinced that Hydrick was performing an old stage magician's trick of concealed breathing to blow the objects around. Hydrick tried, but was unable to move the pages and complained that the Styrofoam had produced static electricity which hampered the telekinetic process. Randi then performed the trick himself successfully. In 1981, while being investigated by journalist Dan Korem and researcher Hugh Aynesworth, Hydrick confessed that everything about his act was a fraud. He had never studied with Kung Fu masters, and the breathing technique he used was one he had developed on his own while in prison. When read, the confession tends to produce more pity than outrage. He comes off as a pathetic and hapless character, ignored all his life, especially by his parents, and his scam a misguided attempt to gain the recognition that he hoped would fill the void in his heart, unless, of course, that even this heartfelt confession was a scam as well. He later spent more time in prison in California for sex crimes. His life story is so full of obfuscations and claimed powers that it is difficult to tell were reality ends and the fantasy begins.

See also: Debunkers.

Further Reading

Korem, Dan. 1997. *The art of profiling: Reading people right the first time*. Richardson, TX: International Focus Press.

INTELLIGENT DESIGN

The current incarnation of creationist thinking, known as intelligent design (ID) is the belief that the living world is such a vastly complex and interwoven continuum that it could only have been designed by some higher metaphysical power and that this condition can be proven by an

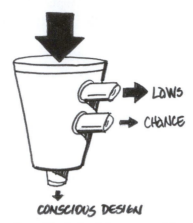

William Demski's design inference machine. Phenomena are loaded in at the top and filtered out the sides, whatever comes out the bottom was intelligently designed by God.

application of modern scientific techniques. Proponents argue that the level of complexity and design seen in the natural world is by itself proof that it could not have come into being through Darwinian natural selection, and that the only possible way to explain the universe is that an intelligent designer purposefully engineered and assembled it. ID believers also argue that their position does not use Biblical references, but is proven by science and as such is a purely empirical and scientific concept.

The belief that God, or the Gods, created the universe goes back to the oldest human civilizations. All cultures around the world, regardless of ethnicity, time period, or religious persuasion, had some type of creation story; intelligent design is only the most recent variation in a long line of such explanations. One thing all creation stories have in common is that they argue life has a reason and purpose: teleology. The formulized notion of tele-

ology goes back to Greek philosophers like Plato and Aristotle. They argued that, if all life has purpose, it must have been designed to fulfill that purpose. If all living things have purpose, then the entire universe must as well, therefore the entire universe was designed with a purpose. That means there must be a designer, a God: in Greek, *theos*. The study of God is called theology. (By definition then, if intelligent design is the search for the grand designer, it is theology not science.) In the Middle Ages, the Christian philosopher Thomas Aquinas adapted teleology to conform to the Christian notion of God, a God who was separate from and superior to the Greek gods and any other gods for that matter.

The precursor to intelligent design was natural theology. This was the idea that there was proof of God's creation in the natural world. It took form in England in the late 1600s, having been born of the increased interest in studying nature for itself and dovetailed into the attempt to use knowledge of the structure and order of the natural world to prove God designed it for a purpose. The best-known theologian to tackle the problem of intelligent design in the late 18th and early 19th centuries was William Paley (1743–1805), who wrote the influential book *Natural Theology* (1802). Paley's novel approach was to use the simple analogy of the pocket watch. The complexity of the watch would make it obvious that it had been designed and constructed for a purpose by a watchmaker. Paley said that nature was like a pocket watch, a complex machine constructed by a designer for a purpose. Therefore, by looking at nature, it was obvious that God had designed it purposefully, therefore, science supported religion.

By the late 20th century the argument took on a new life as ID. The seeds of modern ID theory were planted in the United States in the late 1970s and early 1980s, when it was formed out of the wreckage of the creation science movement combined with the idea that genetic material was a complex coding system. ID proponents wanted—in public, at least—to create an argument that was not theological but still questioned the evolutionary paradigm. The book that became the early movement's holy scripture was Michael Denton's *Evolution: A Theory in Crisis* (1985). A biochemist and medical doctor operating out of the chemistry department of Prince of Wales Hospital in Sydney, Australia, Denton began as an evolutionist, but became increasingly bothered by what he considered a lack of evidence supporting evolution. While he does spend time on the biochemical aspects of his argument, most of the book is given over to arguing that evolution is a socially constructed paradigm rather than one built on empirical evidence. He attacks the narrative quality of evolutionary thinking and argues that its real power comes from being a materialistic and nontheological explanation for life. He accuses evolutionists of being a new priesthood unified in the single-minded pursuit of keeping their idea safe from dissenters.

In 1983, the Christian group Foundation for Thought and Ethics (FTE) began working to publish a creationist book for use in high school science classes. As work progressed on the book, the *Edwards v. Aguillard* case worked its way through the U.S. Supreme Court, which eventually ruled against the inclusion of creationist concepts in public school science classes. The FTE then excised all the overtly creationist references in the manuscript and replaced them with what was thought to be more palatable design language. They also changed the title of the book from *Creation Biology* to *Of Pandas and People* (1989). One of the book's co-authors, Charles Thaxton, is credited with coming up with the term "Intelligent Design," though he admits he borrowed it from engineers who were using it for other areas and applied it to what he was working on. Thaxton was drawn to ID theory by work being done on DNA sequencing and its similarity to the way words and letters are arranged in a book. As language is something that is consciously designed, then DNA must also have been designed. Showing an almost shocking lack of understanding of the mechanics of evolution, the book followed the standard approach of ID publications: show what is wrong with evolution rather than what is right about intelligent design.

The next important work of the ID canon was Philip Johnson's *Darwin on Trial* (1991). Johnson was a Berkeley law professor who wanted to deconstruct evolution along legal lines. Johnson attempted to formulate a purely intellectual and legal argument (though he never really says why Darwin should be on trial). Johnson became a Christian late in life and attacked evolution with all the zeal of a recent convert. He came to believe that modern culture was responsible for the growing secularization of the education system that, in turn, was fostering a general breakdown of American society. At the heart of this breakdown was the teaching of evolution. He claimed his examination of evolutionary mechanics was on its own terms, not in light of any religious framework.

Johnson begins his work with a number of fallacies. In the first paragraph of the first chapter, he shows a fundamental lack of knowledge about current evolution studies. He explains that "scientific orthodoxy" teaches that life evolved "from nonliving matter to simple microorganisms, leading eventually to man." The idea that evolution showed a linear track from simplest to most complex with humans as the epitome of the process went out of favor with scientists decades before Johnson made his remark. This is the equivalent of someone saying that astronomers believe the sun is at the center of the universe; while that was believed early on, modern astronomers understand that the sun is only the center of our specific solar system, not the entire universe.

Johnson also argues that, according to the tenets of science and the words of evolutionists themselves, natural selection as a scientific explanation for life is unscientific. Many evolutionists, he argues, do not believe in evolution because of questions they have raised concerning various parts of the theory. He fails to appreciate that just because a scientist might express doubts about some of the mechanics of evolution, wonder why a particular explanation does not seem to cover every answer, or how a particular fossil has strange aspects to it does not mean he is questioning the overall idea of evolution. Asking tough questions is how science is advanced and how knowledge is gained. This fundamental aspect of intellectual inquiry—doubting, asking questions, and being ready to change one's position—is viewed with deep suspicion by anti-evolutionists and fundamentalist religionists. They tend to prefer knowledge to remain static and unchanging the way holy scripture does. Johnson also suggests that, when scientists do take the time to put forward answers to the uncomfortable questions, anti-evolutionists ask if they are doing it as part of a cover-up. The ID movement is a pernicious establishment flailing away at any and all opponents.

In 1993, Johnson called together a group of men who would become the who's who of the ID movement. Michael Behe, William Dembski, Jonathan Wells, and others met quietly with Johnson at a California seaside resort called Pajaro Dunes to discuss a strategy for putting forward their ideas. They came to the conclusion that they had to be careful to reduce as much of the religious content of their work as possible, in order to make it palatable to a wider audience. The group began arranging a series of public meetings, notably the "Death of Materialism and the Renewal of Culture" conference, which became the basis of the Discovery Institute that, in turn, would go on to become the primary political activist group to get ID into public schools' science curricula. ID and the Discovery Institute were at the heart of all the late-20th- and early-21st-century "Monkey Trials," in particular the Dover, Pennsylvania, case of 2006 in which the state supreme court ruled ID was a religious idea, not science, and therefore could not be taught in a public school science classes.

Intelligent Design proponents continue to argue their work is genuine science, though they cannot point to any ongoing research initiatives to study complexity or any of the other claims they make. Whether or not a deity created the universe, ID theory also oddly undermines religious belief, especially that of their closest allies, the Young Earth Creationists. In order to sound reasonable, ID proponents have said that they do not know who the designer is, and that it could be any deity, or even space aliens. They have also admitted that ID does not preclude the earth being billions of years old, an idea that is deeply offensive to traditional Christian fundamentalists. Moderate theologians have remarked that ID theory is poor theology, while scientists have argued it is poor science. Instead of bringing science and religion together as per the goals of the Discovery Institute, ID theory seems to push them further apart. In the end, ID theory is more a public relations effort than a pursuit of scientific knowledge. That effort is geared toward eliminating materialist science, making it religious, and generating political power for its followers, none of which will help anyone understand how the universe really works.

See also: Creation Science; Irreducible Complexity.

Further Reading

Forrest, Barbara. 2004. *Creationism's Trojan horse: The wedge of intelligent design.* Oxford: Oxford University Press.

Scott, Eugenie. 2004. *Evolution vs. creationism: An introduction.* Westport, CT: Greenwood Press.

IRREDUCIBLE COMPLEXITY

A central theory of the modern creationist notion of intelligent design, irreducible complexity was developed and promoted by authors William Dembski and Michael Behe and posits that living biological entities can only be broken down into their constituent parts. At some point, they can no longer be separated into smaller components, yet these components, while tiny, are still relatively complex structures. This is held as scientific evidence that those biological parts were specifically designed to fit a particular role in the life process; therefore, they can only be accounted for by the presence of and interaction with a higher intelligent designer. Irreversible complexity is a mathematical and statistical, not biological, explanation for how life appears.

The core scientific intellectual of the intelligent design (ID) movement is the philosopher-theologian-mathematician William Dembski, who wrote extensively on complex biological systems and their relation to evolution and argued that microorganisms and their parts, though superficially simple, are still highly complex creatures, too complex to have been formed by chance. Like most leaders of the ID movement, Dembski became an active Christian later in life as he searched for answers to higher philosophical questions. He employs statistics and other mathematical models to show that the odds of life evolving are so astronomically slim as to be impossible. As a result, the only alternative answer to why life is here at all is that it was designed by a higher power.

Dembski argues that there is a "design inference" where the designer occasionally steps in to do things. He concedes that not everything in nature is designed and that some things have happened by chance. To work out the differences, he assembled an algorithmic table to compare and test natural phenomena. If something happens in nature on a regular basis, it is not designed. Dembski sees the creation of life as a very rare event and so the probability that it was designed by a higher intelligence is great. There are two kinds of design patterns in nature. *Specifications* are those whose probabilities for existence eliminate the possibility of having come about by chance. *Fabrications* are those whose probabilities do not eliminate chance. The design inference is a hypothetical filtering device used to determine if a biological element is designed or not. Dembski divides events into three categories: laws, chance, and design. Laws govern events that will always happen under certain circumstances. Chance governs those events that happen as often as they would not happen. Design governs those events that do not fall under the first two categories. Arranged like a funnel with successively finer filters and draining-off ports, the design inference allows for events to be dumped in from the top. Laws are separated out from the stream and shunted off to the side. The few that make it past the first filter are separated out as chance. The minuscule numbers of events that make it through both filters fall out the bottom of the device. It is this relatively tiny number of events trickling out of the bottom of the device that Dembski defines as those that have no other explanation than that they were the product of a conscious design.

The intellectual inspiration for Dembski's work comes from the writings of the Hungarian émigré philosopher Michael Polanyi (1891–1976). Though initially working as a practicing scientist, he became fascinated by the philosophy of science and turned to that field. He had begun to question the basic methodological assumptions of science, especially its grounding in positivism and the belief of its practitioners that what they were doing was objective, value-free, and produced truth. Polanyi's ponderings on the meaning of science resulted in *Personal Knowledge: Towards a Post-Critical Philosophy* (1958). As would be the case of the growing number of postmodernist scholars who argued that, while scientists might be trying to be objective in their work, it was tainted by their personalities and the culture in which they were raised. Their personal feelings interfered with their objectivity. This meant that sci-

ence was not a value-free enterprise after all. His suggestion, like that of the postmodernists, was to accept this personal knowledge aspect of science and move forward. What Polanyi wanted was to create a new epistemology—a way of knowing things—that was more intuitive, less wedded to the pursuit of empirical evidence, and took into account unseen aspects and influences.

Polanyi's work was in a similar vein to Thomas Kuhn, whose widely influential *Structure of Scientific Revolutions* appeared a few years later. Kuhn argued that scientific ideas, or paradigms, are supported not only by facts and empirical evidence but also by the scientific community's attachment to those ideas and their willingness to hold on to them and protect them. Only when there was enough new information, and the community was ready to do so, would one idea replace another in an event Kuhn called a *paradigm shift*.

A central tenet of Polanyi's scheme was the concept of tacit knowledge. This was the intangible, creative aspect of scientific research. Tacit knowledge could not be communicated through books, but only shown or explained by one person to another like in the apprentice system of medieval Europe. He also saw the concept of God as a structure or system related to the structure of life on earth. Intuition and hunches help scientists make the intellectual leaps necessary for knowledge to really move forward and create new models to explain the universe. Polanyi's suggestion was to join formal thinking and experimentation with tacit knowledge. He was looking for a way to examine the universe without having to rely on the intellectual traditions of Enlightenment objectivity and modernity and its rejection of the metaphysical, an attitude

Polanyi saw as problematic and intellectually stifling. He wanted to create an artful blend of science and metaphysics that would give a deeper and more satisfying account of the universe. In *Meaning* (1975), he pondered the place of God in modern science and society and how people search for meaning in a cold society based on objectivity and rationalism, devoid of the mysteries of life. William Dembski imbibed Polanyi's work, as well as that of Kuhn, into his own critique of modern science and society.

Dembski was drawn to Polanyi's work because, to his mind, it undermined the modern notion of what science is all about, something Dembski and the other ID theorists wished to do. For Dembski and other ID theorists, like Phillip Johnson, Michael Behe and Michael Denton, Enlightenment ideals such as positivism, secularism, and the removal of theology from the scientific enterprise are the things that have brought on society's downward slide. Modern science, and evolution in particular, they argued, is a social construction with no relationship to reality. It is simply a model to explain the universe without God or the supernatural. He argued that scientists are not rebels looking for new knowledge, but conservatives protecting an establishment that rejects anything that might question their position. Dembski, like the other members of the ID movement, wants to put God back into science. His plan is to do it using mathematics, probability theory, and genetic sequencing. Using these methods, Dembski is trying to give concrete form to the tacit knowledge that God designed life on earth.

The notion of Enlightenment ideals is crucial to understanding modern anti-evolutionism, the intelligent design movement, and life's irreducible complexity.

The Enlightenment—roughly the 17th and 18th centuries—was the period during which theology began to lose its control over intellectual inquiry in the Western tradition. Enlightenment philosophers were intent on focusing on studying nature by applying methods of reason and empirical fact-gathering over superstition and religious belief. Dembski and his fellows reject the Enlightenment spirit and instead search for ways, like irreducible complexity, to explain scientifically that which is not explainable by science. They are trying to blur the line between science and metaphysics. At a gathering of ID researchers in the late 1990s, called the "Mere Creation Convention," Dembski told his colleagues that what they needed to do was to "redesign" science.

Along with Dembski, the most public defender of design theory and irreducible complexity is Michael Behe, author of *Darwin's Black Box: The Biochemical Challenge to Evolution* (1996). A professor of biochemistry at Lehigh University in Pennsylvania, Behe built upon the notion that the irreducible complexity of a cell showed that life was designed. Irreducible complexity also describes a biological system that consists of connected parts that must all be in place for the larger system to operate. Building upon an old anti-evolutionist argument about what good was half an eye or half a heart while it waited for all the necessary parts to evolve, Behe claimed that, since an eye would not work unless all the constituent parts were present, that alone would prove a designer had to have put the whole thing together at once. Behe used the analogy of the mousetrap, which consists of several interlocking parts that must all be present for the device to function; take one out, and it does not work. If the mousetrap (if it were a living thing) had evolved one piece at a time, what good would it have been until it could perform its function? Evolutionists argue that, if the wooden board of the mousetrap were to evolve first, it would have performed some action. As each piece was added, the device would do what it did more efficiently or might take on other functions altogether. Once enough parts were present, it might start catching mice. Further, as more parts were added, the trap might lose its ability to catch mice and would then perform yet some other function. What many anti-evolutionists often relied on was the notion that evolution was trying to do something specific, like catch mice. Evolution's proponents, on the other hand, argue that the trap was not meant to catch mice; it did so only when the circumstances of structure and environment made it possible. It could just as easily have performed some other function.

The natural theologians of the 19th century had the strength of their convictions to say who they thought the designer was. ID proponents today want to have their work accepted for teaching in the public school system in America under the U.S. Constitution, but they must show it to be secular and not religious. However, Behe and Dembski have both made remarks at private meetings of various religious groups that ID and irreducible complexity are closely aligned with Christian fundamentalism and support the redemptive power of Jesus. Dembski often writes on strictly Christian theological topics. The religious grounding of ID has been acknowledged by Michael Behe in an interview where he said, "Although I find it congenial to think that it's God, others might prefer to think it as an alien—or who knows? An angel, or some satanic force, some new age power." He eliminated any ambiguity at the infamous

Dover, Pennsylvania, evolution trial *Kitzmiller v. Dover,* when he testified that "The designer is, in fact, God."

See also: Intelligent Design.

Further Reading

Denton, Michael. 1986. *Evolution: A theory in crisis.* Chevy Chase, MD: Adler & Adler.

JERSEY DEVIL

This legendary composite creature is said to haunt the forests of New Jersey. More akin to the Mothman than Bigfoot, the Jersey Devil is not normally thought of as a cryptid (an undiscovered animal of genuine biological reality), but as a phantasm. Local folklore describes the Jersey Devil, also called the Leeds Devil, as the creature responsible for terrorizing the Pine Barrens of New Jersey. An enduring local legend, no physical evidence of the creature's existence has ever been produced. All knowledge of the creature and its mostly nocturnal habits are folkloric in nature.

The majority of accounts describe the animal as having the face of a horse and the head of dog with antlers sprouting from the top. Its body is said to resemble that of a kangaroo and it is attached by a long neck. The most notable characteristics are its leathery, bat-like wings, its cloven hooves or pig's feet, fearsome claws, and a thick, reptilian, forked tail. There have been more than 1,000 sightings in southern New Jersey and eastern Pennsylvania alone. While some encounters can be attributed to hoaxes and hysteria, many still cannot be explained.

The oldest and most common story of its origin dates back to 1735. It was deep

A composite of a horse and a bat, the Jersey Devil has been spotted in the Garden State since colonial times but has never been captured or filmed.

within the Pine Barrens (a heavily wooded and secluded region very different from the more popularly known urban and industrial image of the Garden State) where an impoverished Mrs. Leeds learned that she was pregnant for the 13th time. In frustration, she is said to have cursed the child by shouting to the heavens, "I am tired of children! Let this one be a devil." Some say the baby was born deformed. Others claim it was seemingly normal at first, but almost instantaneously transformed into a fiendish creation, right before the mother and her midwives; its body elongated and it sprouted wings and a tail.

Traditional accounts say that Mother Leeds locked the child in the closet for a number of years, until it escaped by flying out through the chimney. A more vio-lent version of the tale states that Mother Leeds, though a Quaker, invited this devil by indulging in sorcery. This version claims that once the baby had transmuted into its new state, it used its large, lizard-like tail to beat, maim, and kill its mother and other women, along with the rest of the Leeds family, before escaping and feeding upon nearby sleeping children. The area was said to be plagued by visitations from then on. The character of Mother Leeds herself may be in part legendary. While remnants of a Leeds family still inhabit central New Jersey, any direct connection to the Mother Leeds of the legend is apocryphal.

The alternate name of the Jersey Devil, the Leeds Devil, is attributed to legends that it was a child of Mother Leeds, but there are other stories that simply claim it is related to the site of its birthplace — Leeds Point, NJ (Atlantic County, formerly Gloucester). According to New Jersey antiquarian Reverend Henry Charlton Beck, it was actually a Mrs. Shourds who gave birth to the creature in a house that was still standing until 1952. Yet another story that regards the birthplace as Leeds Point takes place during the Revolutionary War. It was 1778, when a girl from Leeds Point fell in love with a British soldier during the Battle of Chestnut Neck. It is said that her act of treason was what brought about this curse. Still another take on the origins of the yarn, which did not appear until 1850, describes a young south Jersey girl who refused to help a hungry Gypsy woman. In return, the Gypsy placed a curse upon the girl which was realized with the birth of her first child.

The Jersey Devil has been held responsible for bringing drought and crop failure to the Pine Barrens for years. It has been accused of keeping cows from producing milk and boiling fish in streams. It is a general harbinger of evil, and sightings of it have often prefaced war (it was reportedly spotted in December 1941, right before the attack on Pearl Harbor). There have been a plethora of uniquely disturbing and unnatural events believed to be caused directly by it, such as the sounds of blood-curdling screams and the mass mutilation of pets and farm animals.

The sightings in and around the Pine Barrens in the past 300 years have been attributed to local and famous personalities alike. Of the more well-known people said to have had run-ins with him is American Naval hero Commodore Stephen Decatur. In 1804, he paid a visit to Hanover Iron Works in the Barrens to procure quality cannonballs for use in the fight against the Barbary Coast pirates. It is said that during testing of the artillery, a "bizarre creature" was flying above the firing range. Decatur took aim and shot right through the creature. As many witnesses retold the story, the gaping hole sustained by the Devil did not appear to faze him and he continued to flap along. Joseph Bonaparte, brother of Napoleon and former King of Spain, also saw the Jersey Devil while hunting in the woods near his Bordentown estate in the early 1800s.

A rash of sightings during the third week in January 1909 convinced many that the Jersey Devil was more than just a myth, and boosted its popularity to new heights. As far north as Trenton, east through Pleasantville, down into Wilmington, Delaware, and as far west as Leiperville, Pennsylvania, people were getting glimpses of the glowing "jabberwock," spotting its tracks, and hearing its squawky whistle. Newspapers from all over New Jersey and Pennsylvania reported the stories. Chickens went missing and were found dead with no marks, their bodies scattered in different locations throughout the Delaware Valley. Reputable observers like policemen and court justices also reported sightings. Mrs. Sorbinski of South Camden witnessed her dog being attacked and chased the "flying death" away to rescue her pet.

The Jersey Devil has caused hysteria throughout the southern part of the state resulting in the occasional closure of schools and factories. Fear of it inspired trolley drivers in Trenton and New Brunswick to arm themselves. Reports of the creature have led people to hide behind the locked doors of their homes, and large posses have swept the region. Oddly, any dogs that were to be used to track the "kangaroo horse" all refused to follow the trail. This left hunters to follow the footprints themselves, finding nothing at the end. Farmers set traps and caught nothing besides a few fellow hunters.

A theory was proposed at the time: that it was possibly a prehistoric animal causing the trouble. *Peleosaurus cattelleya* of the Jurassic period could have survived in limestone caves which were lifecycle sustainable beneath the Gulf Stream. The caves were reopened during recent volcanic activity which would have allowed the dinosaurs to be free from their confines. A Smithsonian Institution expert agreed with the underground plausibility, but believed it to be a flying pterodactyl instead. The Brooklyn Meteor Research Society instead believed it to be either a carnivorous marsupial or a fissiped, both of which were also thought to be extinct.

Amidst this hysteria in the early part of the century, hoaxers began to capitalize on the notoriety of the Jersey Devil. Taxidermists fashioned Frankenstein creations, and photographers used trick photography in order to fool people. A promotions man in Philadelphia named Norman Jefferies claimed he had caught the Leeds Devil himself and put it on display. The truth, which came out a few decades later, revealed that Jefferies simply bought a kangaroo, painted it with stripes, and attached bronze, structured rabbit-fur-covered wings to it. He hired a boy to poke it with a sharp stick in order to make it hop up, and this allowed patrons to catch a glimpse.

While this may explain some occurrences, it still does not explain all the unusual tracks, sightings, and animal mutilations that have occurred and continue to the present day. There are many who fastidiously believe in the Devil's existence. One present-day Leeds Point resident, Harry Leeds, claims to be a relative of the Jersey Devil, removed from Mother Leeds by 11 generations. He says the story was passed down his family through the generations, detailing where the events happened, including a personal encounter of his own.

There are creatures of related interest. In Zanzibar, Africa, there is the *opobawa*: a creature described by locals as having an anatomical structure and behaving in ways similar to the Jersey Devil. The *opobawa* has a shorter folkloric life, having been first reported only in the mid-20th century. Closer geographically to New Jersey is the snallygaster of Maryland. It too has anatomy and behavior patterns similar to the Jersey Devil.

The common descriptions of the Jersey Devil and its pseudorelatives the *opobawa* and snallygaster, defy evolutionary mechanics. There are no known vertebrates with hind limbs, fore limbs, and wings (which would effectively give the creature two sets of arms). If some biological and prosaic entity is behind sightings of the creature, it will likely not conform to eyewitness descriptions. Whatever reality lurks behind the Jersey Devil, if any, remains a mystery, but fascination with it remains prevalent throughout the state which gives it its name, so much so that the Jersey Devil has become the namesake of the state's professional hockey team. Ghost-hunting expeditions regularly travel to the towns of Leeds Point and Smithville to wander the lonely rural backroads of the area hoping for an encounter.

See also: Cryptozoology.

Further Reading

McCloy, James F., and Ray Miller. 2005. *The Jersey devil*. New York: Middle Atlantic.

Roberts, Russell. 1995. *Discover the hidden New Jersey*. New York: Rutgers University Press.

Sceurman, Mark, and Mark Moran. 2008. The Jersey devil. *Weird NJ.* (September).

KRANTZ, GROVER SANDERS

American paleo-anthropologist who made a career of trying to prove the anomalous primate known alternately as Sasquatch and Bigfoot was a genuine animal, Krantz (1931–2002) battled resistance from both mainstream scientists and amateur monster hunters for his views and techniques. He attempted to apply the methodologies of physical anthropology to solving the problem of what man-like cryptids were. While he did not originate the idea, he was the primary and most vigorous promoter of the theory that these creatures were the evolutionary descendents of the Asian fossil ape *Gigantopithecus*.

See also: Anomalous Primates; Bigfoot; Cryptozoology.

Grover Krantz was the most outspoken scientific proponent of the existence of Bigfoot. His career suffered for it.

Further Reading

Krantz, Grover. 1992. *Big foot-prints: A scientific inquiry into the reality of Sasquatch.* Boulder, CO: Johnson Books.

Regal, Brian. 2009. Entering dubious realms: Grover Krantz, science and Sasquatch. *Annals of Science* 66 (1): 83–102.

L–O

LAKE MONSTERS

Aquatic cryptids that inhabit inland lakes, rivers, and other freshwater bodies are often labeled "lake monsters." The most famous examples are the Loch Ness Monster, Ogopogo, and Champ. Lake monsters tend to appear on a fairly regular basis, as opposed to their marine cousins the sea monsters and serpents which are often one-off sightings. Individual legends of lake monsters abound from around the world and across time. Large lakes, like the sea, presented a source of wonder and fright to early people. Deep-water bodies, as opposed to shallow lakes and streams, present a vast unknown, as it is difficult to see very far below the surface with the unaided eye. This unknown, dark quality helped produce most monster legends. (Caves and other dark places have the same effect on humans). The 20th-century's great sea monster chronicler was Bernard Heuvelmans (1916–2001), whose work in the field helped inaugurate the modern pursuit known as cryptozoology.

His *On the Track of Unknown Animals* (1958) and *In the Wake of the Sea Serpents* (1968) helped focus attention on lake monsters as well.

The single most famous lake monster is the one said to live in Loch Ness, Scotland. St. Columba, founder of Iona, was the first to notice the creature, during his attempt to convert Scotland to Christianity in the 6th century. The large creature has many descriptions, but is generally described as resembling an eel, dragon, serpent, or dinosaur. It has an estimated length of 30 feet, with a long neck and body covered with moving humps. Less dramatic sightings often describe the creature making a "v-shaped" cut in the water. The cryptic creature is sometimes referred to by its nickname, "Nessie."

Theories of the creature's origin have explained it as a dinosaur, amphibian, serpent, or aquatic lizard. A notable explanation for the Loch Ness Monster is that of a prehistoric, surviving plesiosaur. A plesiosaur is a large, extinct, aquatic reptile distinct from but contemporaneous to the dinosaurs of the Cretaceous period. The waters of Loch Ness, however, are likely too cold to support a reptile of this sort. Ecologically speaking, a carnivore of a plesiosaur's size would not likely survive in the loch. Moreover, sightings may be too rare for the creature to be air breathing, as one can assume the monster would be seen more if it surfaced regularly.

A legend from the medieval era, the Loch Ness Monster was launched to worldwide infamy in 1933 with the publication of the "surgeon's photo." This photo remained the iconic image representing nearly all aquatic monsters until 1994, when it was revealed to have been a staged photo of a model. There have been many expeditions to find the Loch Ness Monster, but only a few fleeting moments of film or photos of ripples in the water left by an unidentified something have been the result. As the loch is a fairly confined space, a number of operations have taken place wherein squadrons of boats carrying underwater sonar equipment have swept the waters, but yielded no better results than sightings by lone observers on the shore. Researchers Robert Rines and Peter Scott searched for Nessie with high-tech sound and photography equipment in the 1970s, and managed

The serpentine head is the most commonly reported anatomical feature of lake monsters.

to take photos of a flipper and gargoyle head-like structure, both of which caused great excitement, but later turned out to have been retouched before publication. As a result, there is no evidence for the creature existing other than eyewitness reports.

Canada, too, has its fair share of aquatic monsters. When Edward Bousefield and Paul LeBlond argued for the existence of the Cadboro Bay monster of British Columbia, Caddy, they put forward evidence that remains of the creature had been found in the past and identified as such. Their primary evidence was a photograph taken in 1937, which was corroborated by period eyewitness reports. In a scientific paper, they proposed to call the creature, more correctly thought of as a sea monster, *Cadborosaurus willsi*. Rines and Scott also put forward the name *Nessiteras rhombopteryx* for the Loch Ness Monster. The other well-known Canadian-U.S. monster is Champ, or Champy, of Lake Champlain. Champ is known mostly through publication of the "Mansi photo," taken in 1977, which has so far defied explanations. However, Lake Champlain is, unlike Loch Ness, a shallow lake. The Mansi photo suggests Champ is a large creature, but the size of the lake calls into question how such a large creature could inhabit the lake at all, let alone go undetected. Another popular North American lake monster is Ogopogo, of Lake Okanagan, British Columbia. There are no photos of Ogopogo, only eyewitness accounts. Skeptics have argued that lake monster sightings are of submerged objects like tree trunks which, from time to time, work their way to the surface, releases of gas from under the lake floor, and misidentifications of non-monstrous animals such as ducks, large fish, or even swimming dogs. With the exception of the Mansi photo, lake monster pictures and films are all of unidentified objects moving across the surface of the water at some distance from the viewer.

One of the most interesting things about lake monsters is that they rarely act like real animals. Lake and river fauna behavior is repetitive. For example, salmon swim upstream to spawn in the same place every year. One need only set up a camera in the right spot, and it is easy to get rolls of clear film showing salmon performing that behavior. Ducks land on lakes and swim across them in the same manner all the time. It would be expected that, if the Loch Ness Monster or Champ were real, they would be easily identified. Cryptozoologists would not need to spend countless hours trying to film them; rather, there would be so much film footage, they would not be considered unusual.

See also: Cryptozoology; Sea Monsters.

Further Reading

Bousfield, E. L., and P. H. LeBlond. 1995. An account of *Cadborosaurus willsi*, new genus, new species, a large aquatic reptile from the Pacific coast of North America. *Amphipacifica* 1 (1): 1–25.

Campbell, Stuart. 1997. *The Loch Ness Monster: The evidence*. New York: Prometheus Books.

Coleman, Loren, et al. 2003. *Field guide to lake monsters, sea serpents, and other mystery denizens of the deep*. New York: Tarcher.

Scott, P., and R. Rines. 1975. Naming the Loch Ness Monster. *Nature* 258:466–68.

LEY LINES

Ley lines are peculiar alignments of ancient ritual sites, stone monuments, and other locations. These alignments are thought to be intentional constructions meant to act as Stone Age highways, with the monuments acting as signposts. The idea of ley lines was first put forward by amateur British archaeologist Alfred Wilkins (1855–1935) in 1921. With a fondness for ancient British stone monuments, Wilkins made a systematic study of their locations and determined that they were not laid down haphazardly, but were carefully arranged and aligned for a cartographic purpose. He laid out his ideas in *Early British Trackways* (1922) and again in *The Old Straight Track* (1925). Similar lines appear at sites in the Americas. Late-20th-century esotericists took Wilkins's ideas and argued ley lines also produced magical effects and occult energies as a result of their alignments and the way they seem to connect Neolithic monuments with presumed magical significance. This association with the occult began with a novel titled *The Goat-Footed God* (1936), by esoteric author Dion Fortune (1890–1946), and has been expanded upon by others. Ley lines have since been connected to dowsing (the lines appear over magical water courses), geomancy (the lines have numerical significance), and, of course, the Nazis (Aryan races also created these lines around Europe). Such lines are now said to be found worldwide; indeed, together they are said to make up a planetary grid system that taps into cosmic energies.

See also: Geomancy.

Further Reading

Cathie, Bruce L. 1997. *The energy grid: Harmonic 695: The pulse of the universe*. Kempton, IL: Adventures Unlimited Press.

Williamson, Tom. 1983. *Ley lines in question*. n.p.: World's Work.

LIVING DINOSAURS

Persistent belief that dinosaurs lived or continue to live contemporaneously with humans. As with many of the subjects in this book, there are pseudoscience living dinosaurs and genuine living dinosaurs. The genuine living dinosaurs are birds, the evolutionary descendants of several strains of smaller theropod dinosaurs. The evidence of birds' ancestry is considerable and, while scientists do disagree over some aspects of the bird-dinosaur link, it has been well established in general. Evidence for pseudoscience living dinosaurs, put forward by cryptozoologists and creationists, is far harder to come by, and some of it is outlandishly incorrect.

The classic dinosaurs, models of which can be found in any kid's toy box, became extinct about 65 million years ago. Their fossil remains have been found around the world and in some abundance. Because of their natural strangeness and sometimes-unusual size, they are favorites of museumgoers everywhere. Current thinking holds that some major ecological catastrophe—possibly a meteor strike—or combination of changing environmental conditions pushed them into extinction. By the time this occurred, however, several lines of smaller, raptor-type species had begun to evolve feathers and wings. Species such as *Proavis* and the better-known *Archaeopteryx* had appeared. Some of these species went on to become modern birds that managed to survive the great extinction and continued on to the present.

The pseudoscience living dinosaurs come in two categories put forward by two different special interest groups: cryptozoologists and creationists. The more reasonable of the two are the relic dinosaurs of cryptozoology. These creatures are based upon legends, especially from Africa, of monstrous animals that seem to fit descriptions of dinosaurs. The best known is the Congo's *Mokele-Mbembe,* which is said to resemble a sauropod and to inhabit the region around Lake Tele. As with so much about cryptozoology, Bernard Heuvelmans and Ivan Sanderson were among the first to engage with these legends. In a series of books written in the 1940s and 1950s, Willy Ley (1906–1969) also addressed the topic of living dinosaurs. Ley was a German paleontologist turned aerospace engineer and colleague of Werner Von Braun. Unlike Braun, Ley was unwilling to put his expertise to work for the Nazis, and so he left the country in 1935 and came to the United States to pursue a career, like Ivan Sanderson and Bernard Heuvelmans, as a science writer. He dabbled in writings on unusual animal life that were well within the parameters of what would be called cryptozoology, but what he called "romantic zoology." In 1949, he published *Do Prehistoric Monsters Still Exist?* in which he discussed the legend of *Mokele-Mbembe,* and began by stating that "Dinosaurs may still roam the unexplored jungles of Africa!" Ley collected reams of source material on pygmies, giants, primates, and anomalous animals as references for his writing, and even copied entire chapters of Heuvelmans's *On the Track of Unknown Animals* for his own reference. These included materials on medieval monstrous races, legends of little people, and living dinosaurs of Africa. In 1951, George Sarton, the editor of the influential journal *Isis* and one of the founding fathers of the modern field of the history of science, wrote to Ley to congratulate him on the publication of *Lungfish, Dodo and the Unicorn.* He then went on to relate to Ley a humorous story of dinosaurs in Africa. So pervasive was the belief that dinosaurs still roamed the region it seems that the Governor General of The Congo put out an edict during WWI stating that any dinosaurs traveling at night were required to carry warning lights.

The more problematic category of living dinosaurs is put forward by creation scientists and other anti-evolutionists. Traditional Young Earth Creationists believe a literal interpretation of the Bible and, as such, hold that the earth is no more than 10,000 years old. They reject the geological time scale and fossil evidence for an ancient earth. They use a quirky interpretation of the fossil record to argue that humans and dinosaurs were contemporaneous in the recent past. A cornerstone of this argument are the man tracks of Texas, first discovered in 1917 along the banks of the Paluxy River outside the town of Glen Rose. Many of the tracks are clearly those of dinosaurs, but a few take on odd shapes because of the distortion caused by the animals walking through what at the time was wet mud. It is these tracks that creationists argue are human footprints. Roland T. Bird (1899–1978), the American Museum of Natural History excavator who did the first extensive work at Glen Rose in the 1930s, tried to convince people that the prints were nothing more than deformed dinosaur tracks. Despite the reality of what these tracks are, creationists have built an entire industry around them and similar examples they see as supportive of their anti-evolutionary thinking.

One of the delicious ironies of the living dinosaur concept is that if cryptozoologists ever prove that creatures like *Mokele-Mbembe* exist and that they are genuine dinosaurs, it would support creationists' contentions of human-dinosaur relations. It would only be a partial victory, however, because, if dinosaurs did make it past the great extinction and into the human era, it still would not undermine the geologic time scale or evolution, but in fact support it.

See also: Anomalous Fossils; Creation Science; Cryptozoology; *Mokele-Mbembe*.

Further Reading

Ley, Willy. 1949. Do prehistoric monsters still exist? *Mechanix Illustrated* (Feb): 80–144.

Ley, Willy. 1959. *Willy Ley's exotic zoology.* New York: Viking Press.

MACKAY, CHARLES

British author of *Extraordinary Popular Delusions and the Madness of Crowds* (1841), Charles Mackay was one of the first modern writers to address pseudoscience and fringe belief. Chapters in the book include examinations of timeless delusions like fortune-telling, alchemy, prophesy, and witch crazes, as well as those specific to his time, such as tulipomania and the Mississippi Scheme. Mackay (1814–1889) saw interest in these topics as large scale delusions where entire societies and cultures fell prey to hysteria over ideas that would normally be of interest only to a few or be ignored altogether. In his introduction, Mackay said, "in the reading of the history of nations, we find that like individuals, they have their whims and their peculiarities." He wanted to know why people who normally behaved in rational and sober ways would suddenly abandon that behavior and embrace dubious realms with such gusto; as a result, *Extraordinary Popular Delusions* can be read as an early study of mass psychology. Mackay was a journalist, born in Scotland, and was named editor of the *Illustrated London News* in 1852. He also published poetry and songs that were popular. His book is still read by students of both economics and pseudo-science.

METAPHYSICS

The intellectual pursuit that investigates those topics which science is unequipped to investigate. From the Greek meaning beyond (*meta*) the physical, metaphysics deals with those ideas and concepts of intangible and often spiritual qualities. The studies of the meaning of life or the presence or intention of God are classic metaphysical topics. More philosophical than experimental, more transcendent than concrete, metaphysics, unlike science, can take into account the supernatural without any intellectual difficulty. While there were undoubtedly individuals who considered such questions before, the first self-styled metaphysicians were the philosophers of ancient Greece like Aristotle and Plato. Metaphysics encompasses such ideas as ontology (the study of the nature of reality) and epistemology (the study of the nature of knowledge). One of the first questions tackled by ancient metaphysics was the nature of being, as people struggled to account for themselves and their place in the universe. In the most simplistic terms, metaphysics can be said to be the pursuit of meaning of being. Initially, metaphysics was considered a normal part of what would later be called science (knowledge). A more empirically based aspect of the discipline came to be known as natural theology. Popular in the 18th and 19th centuries, natural theology was a concept supporters claimed used a materialist study of nature to help prove the existence of a divine element: God (a poor modern relation to this philosophy is the intelligent design theory). By the 20th century metaphysics and science had been separated into distinct spheres. Science supporters argued that science must be able to generate empirical evidence for something. It must be able to measure, weigh, test, and physically examine its subjects. Scientists claim they have no tools or techniques for examining metaphysical concepts; therefore the two disciplines must remain separate for both to remain valid pursuits. A mixing of science and metaphysics would dilute and undermine both.

Further Reading

Kim, J., and Ernest Sosa, eds. 2000. *A companion to metaphysics.* Malden, MA: Blackwell.

Loux, Michael. 2006. *Metaphysics: A contemporary introduction.* New York: Routledge.

Lowe, E. J. 2002. *A survey of metaphysics.* Oxford: Oxford University Press.

Williams, Michael. 2001. *Problems of knowledge: A critical introduction to epistemology.* Oxford: Oxford University Press.

MIRACLES AND MIRACLE CURES

Miracles and miracle cures are inexplicable events attributed to divine intervention. Often associated with good fortune and rewards for proper behavior, miracles are seen as the result of specific communications with a deity or deities through sacrificial offerings or ritualized poetry—prayers—or simple, crudely worded requests in which a supplicant asks for assistance with an especially difficult problem which has resisted human reconciliation. The most commonly requested miracles are those related to cures of illnesses or injuries that have resisted standard medical practice. Miracles can also come in the form of found objects, avoided dangerous mishaps, or otherwise beneficial incidents the recipient cannot account for through any other means. All cultures and religions have examples of such divine interventions. Miracle cures, also known as faith healing if a religious intermediary is present, are those that defy medical science. They come most often in the form of a remission of some terminal disease or sudden recovery.

There are a number of specific locations that are associated with miracles and miracle cures. One of the best-known locations for miracle cures in Lourdes, France, near the Spanish border. Tradition holds that, in 1858, the Virgin Mary (the mother of Jesus, also known as the Blessed Virgin Mary or BVM) appeared to a peasant girl named Bernadette Soubirous (1844–1879). Investigations by local Catholic Church officials found her not to be falsifying her account: she was eventually canonized in 1933. The spot where the BVM appeared quickly became a pilgrimage site and throngs of worshippers began to arrive to drink the water from a spring that appeared—also miraculously—at the spot. Pilgrims began to report being cured of various diseases and other maladies as well. This only increased the flow of tourists and subsequently the number of reported miracle cures. The appearance of the BVM often turns a location into a miracle cure site. Like Lourdes, Fatima is another place where the BVM appeared to three peasant children, in 1917. Located in central Portugal, Fatima has also become a pilgrimage site said to be imbued with miraculous healing powers.

Along with spontaneous miracle cures that occur when a sufferer asks God directly for such aid, there is faith healing. Christian faith healing finds its mandate in the New Testament, as Jesus is said to have cured people by the laying on of hands. Others throughout the ages have also adopted the practice of faith healing. This is where a religious intermediary—a pastor or other religious leader—channels God's power to a sufferer. While faith healing occurs in many religions, it is most often associated with Pentecostal and evangelical protestant Christianity. In this milieu, faith healing is performed in a church in front of an audience and tends to be more dramatic and theatrical than the cures granted directly by God at Lourdes or Fatima. Faith healers rarely take public credit for the healing; instead saying they are simply acting as a conduit for divine healing power. Faith healers also sometimes perform psychic surgery, as seen in the case of Brazil's John of God (JOG), who with no medical training performs "surgery" on patients. Using his hands and occasionally a scalpel, but no anesthetic, JOG removes "tumors" and other growths from the patient's bodies. The patients then claim to be cured.

Faith healers come in the form of two characters, the humble peasant, like John of God, and the flamboyant performer, such as American televangelist Peter Popoff. Both styles of healers employ the laying on of hands, as Jesus did. In the 1980s, Popoff had a large and prosperous faith healing program that drew large crowds to his televised church, where he regularly claimed to cure people of various ailments. Paranormal investigator James Randi showed Popoff to be a fake and a con artist, ending his career for a time.

Miracles and miracle cures are not restricted to self-appointed religious holy men. There are miracle cures associated with modern high tech medicine as well. Man-made wonder cures have a long history. Some of the most flamboyant were those that permeated American culture in the 19th century. Now often referred to as "snake oil," these cures promised much and delivered little. These concoctions were also known as "patent medicines" because U.S. law at the time allowed them to be officially patented, though the patent process was merely a paperwork shuffle with little or no government oversight. These medicines were usually water mixed with alcohol so that the user would feel a brief respite from his problems. American medical history also has a long tradition of amateur doctors, known as "irregulars," who promoted the anti-intellectual idea that they were undermining an evil and uncaring medical establishment to allow people to get the cures they needed; as a result of these claims, they were quite popular. This anti-medical train of thought still underlies the alternative medical movement in the 21st century. In Delhi, India, in late 2005, a medical clinic doctor named Geeta Shroff claimed her work on stem cells—specialized human cells that can grow into a wide range of tissues and organs and which, even under strict scientific protocols, are controversial—had produced a cure for a variety of illnesses and afflictions, including cerebral palsy. Though Indian law allows doctors to use untested methods on cases considered terminal or incurable, Shroff refused to give skeptical colleagues any explanation for how her miracle cures worked. In the city of Hyderabad, India, a family of herbalists has produced a remedy for asthma that users have considered miraculous for over a century and a half. It is given out free to sufferers twice a year based upon astrological logic. The give-away draws tens of thousands of sufferers to the city.

Pseudo medical miracle cures come in a wide range of forms. They usually have been subject to no scientific or clinical tests, are promoted by charismatic pitchmen or personalities, are sold through unconventional means, and are said to cure a wide variety of problems. Medical miracle cures have secret ingredients and purveyors claim they are battling with the medical establishment who wants to keep them quiet.

One religion that is often accused of being antidoctor is Christian Science. Founded in the later part of the 19th century by charismatic healer Mary Baker Eddy (1821–1910), Christian Science is a theology that argues disease, illness, injury, and even death are illusory and the result of poor spiritual health. Church members believe that prayer, faith, and the right Christian thinking can cure without the need for doctors or drugs. Eddy laid out her theology as a practical "science" in *Science and Health* (1875), the book that serves as the central text, along with the Bible, of the Christian Science movement. While not proscribing the use of modern medicine, Christian Science theology steers followers away from it. Eddy herself had a deep suspicion of doctors, but not, in her quirky way of thinking, of dentists. Today, Christian Science takes a pragmatic approach to doctors and illness. While church leaders still prefer Christian Science practitioners—those specially trained to heal others with prayer—to doctors —and support church members in legal suits over the issue, they do not, as a rule, ostracize members who opt to consult the medical establishment.

Miracles are as pervasive in modern society as they have ever been, and prayer is cited as a catalyst to help bring on such events. While science has attempted to learn if prayer is a catalyst to the improvement of health, it cannot determine if miracles exist. Francis Galton (1822–1911), the founder of the pseudoscience of eugenics, conducted the first known study of the effect of prayer on illness. He concluded that prayer did not have an effect upon individuals' health. Numerous studies followed. In 2001, the prestigious Mayo Clinic conducted a study of prayer and healing and also found no correlation between being prayed for and an improvement in health. While prayer might make a person feel good, it has not been shown to cure or heal anything but perhaps a broken heart. Many miracles and miracle cures have been investigated by doctors (for a healing to be categorized as "miraculous" at Lourdes or Fatima, doctors must be able to show that there was no possible medical explanation for the cure) and normally show some medical reason for the cure to have occurred. Faith healing cures have been investigated and it has been found that they most often relapse within a few days, weeks, or months. Those few anomalous cases where no medical explanation can be found do not mean they were miracles. Miracles, especially the avoidance of catastrophe claims, only seem miraculous after the fact and with a good bit of reinterpretation of events, though doctors do accept that spontaneous cures do occur and that explanations elude them. Again, just because no explanation can be found at the time of a cure, it does not mean it was a miracle, it only means that we still have a lot to learn about how the human body works.

Miracles of all kinds raise various intellectual as well as scientific problems, as they cannot be checked by science. By their very nature, they are metaphysical events beyond the realm of complete quantification and investigation. In addition, miracles suggest a kind of hit or miss and capricious divinity. If two persons need or pray for a miracle and only one gets it, does that mean the nonrecipient is unworthy? If a good person, someone pure of motive, prays for a miracle and does not get it, does that mean that God was not paying attention? Miracles seem to pose more theological problems than answers.

See also: Alternative Medicine; Debunkers; Hydrick, James Alan.

Further Reading

Byrd, R. C. 1988. Positive therapeutic effects of intercessory prayer in a coronary care unit population. *Southern Medical Journal* 81: 826–29.

Gill, Gillian. 1998. *Mary Baker Eddy*. Reading, MA: Perseus.

Lourdes. 1911. *The Catholic Encyclopedia*, Vol. 12. New York: Robert Appleton Company.

Nickell, Joe. 1998. *Looking for a miracle: Weeping icons, relics, stigmata, visions & healing cures*. New York: Prometheus Books.

MODERN NEANDERTHALS

The early human-like hominids known as Neanderthals were so much like modern humans that they, more than any other prehuman species, have been the focus of much fascination. Their fossils have been found across Europe, the Middle East, and North Africa. Because of the harsh environments they inhabited, they were far more robust than modern humans and had naturally powerful physiques. The most distinctive aspects of their anatomy were their flattish heads and heavy brow ridges. They first appear in the fossil record about 200,000 years ago, and died out about 50,000 years ago. Along with mainstream paleo-anthropological theories about who the Neanderthals were, how they lived, and what their relationship to modern people was, they have also been used as the central characters in less mainstream thinking by cryptozoologists and anti-evolutionists. The first and more reasonable of these alternative histories is the notion that small groups of Neanderthals continued to live into the modern era and are the basis of accounts for the sightings of Bigfoot-like creatures, especially in Asia where they are known as *Almasti*. The other is that the Neanderthal fossils scientists claim are the remains of ancient humans who died out at least 50,000 years ago are, in fact, the remains of modern people who were described in the Bible as having lived to immensely old ages.

The idea that relict Neanderthals are at the heart of Asian Bigfoot sightings was the brainchild of Russian historian Boris Porshnev (1905–1972). He put forward this theory in the late 1960s and became so associated with the idea that it is also known as the "Porshnev Theory." His ideas were picked up and promoted by some Western anomalous primate researchers, most notably the founders of cryptozoology, Bernard Heuvelmans and Ivan Sanderson. There are several cases of sightings of *Almasti* by modern observers in the rural regions of the Russian hinterlands. The best-known are stories of a woman named Zana said to be living in an isolated Russian village who was one of these relicts. She was said to have borne children with local men, so her offspring were the hybrid products of a union between a modern man and a Neanderthal female. Porshnev first described this case in the early 1970s.

One of Porshnev's colleagues, Alexander Mashkovtsev, heard rumors of a woman in the Abkhazia region of the Caucasus Mountains who was a "wild woman" captured by peasants in the late 19th century. She was subsequently sold or passed around until she came into the possession of a nobleman named Genada who took her to his estate near the town of Tkhina. Her life at the estate was not very encouraging; she was kept outside in a series of enclosures and was often shackled; peasants threw rocks at her and generally abused her. She was thought more animal than human, and was said to have an ape-like face. She never learned to speak, but her bestial nature was apparently not a barrier for some men, possibly even the nobleman Genada, because Zana gave birth numerous times over the course of her sad life of captivity. Some of the children survived. Her youngest, a boy named Khwit, passed away as late as 1954, when he was in his 60s or 70s. Porshnev tried unsuccessfully to find Zana's grave in order to examine her remains to determine if she was a relict Neanderthal. After his death, Dmitri Bayanov took up the case and managed to find Khwit's grave. While the skull did seem to show Neanderthal characteristics, a DNA test showed he was entirely human. Bayanov felt that Khwit was a good candidate for being a human-Neanderthal hybrid, but this seems unlikely. Zana was likely an unfortunate feral human, possibly mentally handicapped or hopelessly traumatized, who, because of her wild appearance, was treated as an animal. As of this date, no proof of relict hominids has yet been discovered.

The other Neanderthal theory, one that is unconnected to the Porshnev Theory, holds that Neanderthals were recent people, no different from modern humans except that they could live for hundreds of years and were immune to most human diseases and other maladies. This theory's most vocal proponent is New Jersey dentist and Biblical literalist Jack Cuozzo. His 1998 book *Buried Alive* tells of his research into Neanderthal dentition. Following the unwritten protocols of hidden history, Cuozzo claims the scientific establishment is trying to keep this information secret. Subtitled with great pathos *Hidden, Suffocating from the Pain of a Story Left Untold* . . ., the book is part science text, part adventure novel, and part religious sermon, but, despite the book's subtitle, Cuozzo never explains why the Neanderthals would suffer from the pain of a supposedly untold story.

Cuozzo based his thesis upon a survey of Neanderthal teeth in various museum collections around the world. Though he claims that the forces of evolution were determined to stop him from studying the material, in his introduction he thanks many scientists and museum staff for graciously granting him access to their prized specimens. He believes a dental expert could see things in the fossil teeth others had missed. He argues that "no person with a creationist world view . . . has ever penetrated behind the evolutionist's lines to study their fossils with X-Ray equipment."

Cuozzo argues the Neanderthals were unique in that they had physiques and longevity far in excess of modern humans. They were impervious to many of the medical ills that plague us, including radiation poisoning. He found that Neanderthal teeth are modern and that modern people are not evolutionary advances over earlier Neanderthal physiology, but rather the results of devolution and degeneration from the ancient ideal. Cuozzo argues that Neanderthal anatomy, far from supporting the evolutionary view of human development, proves the Bible correct. Having appeared just after Noah's flood, the Neanderthals lived their extended lives, in some cases living as long as 400 years, but eventually began to degenerate, as the Bible says they did. Modern people, on the other hand, have lost all the advantages the Neanderthals had acquired from God and are continuing to be run down by the process of devolution. Cuozzo never explains why God created this separate human race, nor does he approach the apparent flaw in his scenario that suggests modern people evolved from the Neanderthals.

In addition to the dental evidence showing the modernity of the Neanderthals, Cuozzo points to other anomalous evidence. He takes the skull of a Neanderthal commonly called "Rhodesia Man" as a case in point. Found in the early part of the century in Zimbabwe, Rhodesia Man—also known as the Broken Hill Skull—contains a hole that anthropologists have puzzled over. Most think it the scar of an abscess or lesion the man suffered in life. Others have asked whether it was the result of trepanning, the practice of drilling holes in the skull that is known to have been performed by early historic people. There are many examples of these successful early brain operations. Cuozzo dismisses all these explanations, claiming instead that the skull of this supposed ancient human ancestor was pierced by something else entirely; to Cuozzo, the wound looks like it was made by a modern bullet. Thus, this individual graced by God to live hundreds of years, impervious to those ills that plague humans today, was, Cuozzo claims, dispatched by a shot through the head.

Cuozzo believes God Himself was guiding him on his mission and was, through His diligent efforts, thwarting the work of the scientific establishment. Cuozzo's proof is that he claims he was being chased by evolution's dark agents. At one point during a trip to Paris to photograph Neanderthal skulls, he and his family stopped at a pizza parlor. Cuozzo was horrified to see a man who had been following them sitting just a few tables away. Cuozzo scooped up his family and raced wildly back to their apartment through the rainy Parisian night and barricaded the door. Somehow the Cuozzos managed to make it safely back to the United States with their evidence intact.

Astounded by the gravity and implications of his findings, Cuozzo attempted to get specialists to view his work. He claims that the well-known evolutionist Christopher Stringer was intrigued and worried by it. Cuozzo first contacted Stringer with his results in 1986 and jolted him out of his arrogant complacency. According to Cuozzo, Stringer was so taken by his data that he quickly published a paper refuting in advance anything Cuozzo might publish. According to Cuozzo, Stringer and his coauthors argued that tooth-eruption rates for Neanderthals were the same as for modern humans. Cuozzo argues that slower than modern tooth-eruption rates were an indication that Neanderthals matured slowly and lived longer than modern people. He is convinced that Stringer was covering himself, and evolution science, by preempting his work. Like a true missionary, Cuozzo will not be deterred from his sacred task, despite the efforts of the establishment to keep him from publishing his book reinforcing a belief in a supernatural God.

The evidence that Neanderthals were ancient people is overwhelming. There are disputes between scientists over just how the Neanderthals fit the human evolutionary map, however. These disputes revolve around whether or not the Neanderthals were a line to humans that went off into an evolutionary dead end by themselves, leaving no descendents, or whether they were in the direct line of modern humans. Recent studies have shown that humans do share some traits with Neanderthal DNA.

See also: Anomalous Primates; Cryptozoology; Hidden History.

Further Reading

Bayanov, Dmitri. 1996. *In the footsteps of the Russian snowman*. Moscow: Crypto-Logos.

Cuozzo, Jack. 1998. *Buried alive*. Green Forest, AR: Master Books.

Heuvelmans, Bernard. 1958. *On the track of unknown animals*. Paris: Plon.

Porshnev, Boris. 1969. Troglodytidy i gominidy v sistematike i evolutsii vysshikh primatov [The Troglodytidae and the Hominidae in the Taxonomy and evolution of higher primates], *Doklady Akademii Nauk SSSR* [*Current Anthropology*], 188:1. 15:449, 1974:450.

MOKELE-MBEMBE

African cryptid said to be a relic dinosaur. Reported in the west African countries of Congo, Cameroon, and Gabon, general descriptions of the creature by local people are similar to those of a sauropod dinosaur. The pygmies of the Lake Tele region claim to have seen and hunted it. Western reports go back to the mid-18th century. Throughout the 20th century, a series of expeditions have been mounted in hopes of locating the animal, but have been without success. Something always seems to happen just before a sighting so that cameras are never ready to be used, or after a sighting, notes, drawings, photographs, and other proof are somehow lost. The name *mokele-mbembe* may stem from monster-hunter Ivan Sanderson's attempt to spell the word local people told

Mokele-Mbembe *is a supposed living dinosaur that roams the Congo River area of Africa.*

him it was called. He spelled it *m'koo m'bemboo*. Inhabitants of the region where the cryptid has been seen have numerous names for it and several different descriptions of it.

See also: Cryptozoology; Living Dinosaurs.

Further Reading

Nugent, Rory. 1993. *Drums along the Congo: On the trail of* Mokele-Mbembe*, the last living dinosaur.* New York: Houghton Mifflin.

MONSTROUS BIRTHS

Medieval and Renaissance term given to birth defects of both humans and animals. Deformities included conjoined twins, multiple limbs and heads, or any profoundly altered conditions. Prior to the modern understanding of genetics and teratology (birth defects), medieval people had little choice but to ascribe supernatural causes to such births. The word "monster" comes from the Latin *monstrum* meaning "omen." The term also suggests a warning or prodigy. The appearance of such a birth was often seen not only as a bad sign for the family, but also a warning of bad things to come for the local community. Scholarly treatises were written on the subject, as the learned tried to explain unusual conditions. One of the best known of these works was by pioneering French surgeon and anatomist Ambroise Paré (1510–1590) whose *Des Monstres et Prodogies* (1573) included both a synthesis of other writers' works with a good bit of original thought. Paré separated all such occurrences into two categories. Monsters were naturally appearing deformities, while prodigies were more aligned with the supernatural. A great amount of popular literature grew around monstrous births and focused on their religious and moral meaning as their puerile exploitation typically generated healthy sales.

Well into the 18th century, monstrous births were used by Christian clergy and self-appointed moralists as examples of the consequences of human folly and godlessness, and religious pundits used monstrous births as vehicles to address issues of morality and behavior. Preachers and pamphleteers used both genuine and fictional cases to prove their points, which usually centered on female immorality and wantonness. Monstrous births were said to be the result of the mother's (or only rarely, the father's) poor behavior. Specific deformities were said to denote specific failures of moral judgment. Not only would sexual promiscuity result in a woman giving birth to a deformed baby, independent thinking on the part of a woman, resistance to marital subservience, and the wearing of ostentatious clothing could produce them as well. Women, therefore, were taught to be good and subservient wives and pious Christians, lest they give birth to a horror.

There was more to learn from monstrous births than just lessons regarding individual behavior. In post-Reformation Europe, for example, preachers used monstrous births to generate fear of Catholics. An animal's monstrous birth that gained wide acclaim was the "monk calf," which purportedly had a deformity of the head that made it look like the hood of a Catholic monk's habit. Martin Luther himself made reference to this creature in his railing against the Catholic Church. Pamphleteers also made extensive use of illustrations in their work, so that the illiterate or barely literate—the bulk of their constituency—could see the folly as well as read about it. This also fed the human appetite for bizarre imagery.

By the 16th century, medical and scientific researchers were examining monstrous births as a way to learn about human generation. The rise in the use of dissection as employed by Vesalius, Da Capri, and others contributed to the understanding of human variation and such cases were becoming more a topic of scholarly study. Many contemporary authors, however, still saw monstrous births as symbols of disorder and chaos in the universe, as something against God's will.

Alan W. Bates, in *Emblematic Monsters* (2005), argues that monstrous births began to be seen as fitting into the "great chain of being." This early idea about the ordering and relationship of different life forms was a step in the direction of modern understandings of how all life on earth is related. Bates argues that monstrous births were used by some scholars to argue that the chain did indeed exist and that, far from being a symbol of disorder, monstrous births had their place in the wider scheme of life on earth and represented a divine order. Some Renaissance and Enlightenment authors saw monstrous births not as aberrations but as part of nature. The evolution of early studies of monstrous births as prodigies into the modern science of teratology is often put down as the early steps in the inexorable march from superstition to enlightened science. While this certainly contributed to the change in attitudes, some modern writers point to the growing separation of popular pamphleteering and scholarly literature as a deciding factor in making the study of monstrous births more reasoned and less sensational.

There were a few cases of monstrous births that did not portend evil, but were instead thought to bring good fortune. In the village of Biddenden, Kent, England conjoined girls were born in 1100. Due to a lack of medical understanding of the condition or the general repugnance of such cases during this period, monstrous births rarely lived for very long. Occasionally, they were intentionally destroyed. The Biddenden Maids, Mary and Elizabeth Chulkhurst (1100–1134), as these girls were known, were conjoined but were otherwise not deformed. Modern analysis of the story suggests they were pygopagus twins: connected at the lower spine and pelvis. They lived to adulthood and could manage quite well on their own. They were born into a wealthy family, so, despite their condition and ominous arrival (legend says they were born on the day of the assassination of King William Rufus), they prospered. Upon their deaths, they left their wealth to a local church and decreed that their money be used to feed the poor every Easter. In addition, tiny biscuits in their image were also handed out. They are still remembered in the town. In Delhi, India in 2008, conjoined twins were born and the local culture embraced and revered them as a good omen because of their resemblance to multi-armed, multilegged Hindu gods.

By the 19th century, monstrous births had lost most of their religious baggage, though not their sensational allure. Possibly the most famous monstrous birth was that of Chang and Eng (1811–1874), who were born in Siam (now Thailand) and became known as the "Siamese Twins." Similar in appearance to the Biddenden Maids, the Siamese Twins were brought to the West and put on tour. Their popu-

The Monk Calf was used by Martin Luther to show the perfidy of the Catholic Church.

larity was such that they retired to live in the United States as rather wealthy landowners and gave their name to the entire group of conjoined twins. The other well-known monstrous birth of the 19th century was that of the Italian Tocci brothers, Giacomo and Giovanni Batista (1875-?). With only two legs between them, they also did well financially on tour and were able to retire to a secluded villa. Their lives are shrouded in ambiguity so that many of the details of their lives, including their birth and death dates, are obscure.

Due to a better modern understanding of what monstrous births are through the science of teratology (a term first used in the 1960s), such births have a better chance of being separated, and those that cannot have an improved life span. It is understood today that monstrous births are not the result of the poor behavior of the mother, but rather are genetic defects over which the individual has no control. Today when they appear, the interest surrounding them is as focused on the technique of separating them as it is on deformity itself. In fact, the term "monstrous birth" is no longer employed and is considered insulting, in the way the term "retarded" is no longer used to indicate an individual with a learning disability.

Further Reading

Bates, A. W. 2005. *Emblematic monsters: Unnatural conceptions and deformed births in early modern Europe.* New York: Rodopi.

Bondeson, Jan. 1992. The Biddenden maids: A curious chapter in the history of conjoined twins, *Journal of the Royal Society of Medicine* 85 (4): 217–21.

Bondeson, Jan. 2000. *The two headed boy and other medical marvels.* Ithaca, NY: Cornell University Press.

Crawford Julie. 2005. *Marvelous Protestantism: Monstrous births in Post-Reformation England.* Baltimore, MD: Johns Hopkins University Press.

Wallace, Amy, and Irving Wallace. 1978. *The two: The story of the original Siamese twins.* New York: Simon & Schuster.

MONSTROUS RACES

The medieval and Renaissance European designation for foreign peoples thought to inhabit the far corners of the earth. These were mostly fictitious groups or actual people of whom Europeans had a distorted and fantastical image. The Greeks, like Aristotle, Plato, and others, described monstrous people and animals from north Africa, Ethiopia, and India in their writings. The word "monster" derives from the Latin word *monstrum* meaning an omen, warning, or prodigy. The most significant Roman author on the subject was Pliny the Elder (23–79 AD) whose *Natural History* became the com-

mon source work for later European authors. Pliny himself based his writings on works that predated him. Pliny's work on the subject, along with other authors of antiquity, was compiled in the manuscript known as *The Wonders of the East,* copies of which were produced between 970 and 1150 AD. Little firsthand research into the nature of these races was conducted, and only the vaguest attempts were made to verify accounts. Descriptions were largely based upon travelers' memoirs, which were themselves often distorted and sensationalized and put together as remembrances of yet other travelers' tales. Manuscript illuminators then took liberties with how the monstrous races were depicted.

Every culture imagines itself against "the other," and a self-image is generated in this way. Europeans were no different. By the Middle Ages, Christians in Europe were comparing themselves to other groups around the world in order to differentiate themselves. They cobbled together a view of foreign, non-Christian people that was spectacularly inaccurate and so different from themselves that they thought of the others as monstrous. These were people only known or understood from legend, classical mythology, travelers' tales, and half-remembered stories. Representations of the outside world and its inhabitants included giants, long-haired fish-eating people, the *Homodubi* (half man–half donkey), and other fantastic creatures. Some of the most popular were the *Blemmyae* (men with their faces in their chests), the *Cynocephali* (dog-headed men), the *Panotii* (big-eared men), and the *Sciopodii* (men with one giant foot). Certain visual cues and motifs were employed to show the primitiveness of monstrous people; they carried clubs as a symbol of their barbarous state, were hairy and unkempt, and were almost always naked. All these were signs of their cultural, intellectual, and spiritual backwardness, inferiority, and dangerousness.

Images of the monstrous races merged with European legends of barbarians and wild men. Beginning in the 400s, the "civilized" societies of the old Roman Empire were ravaged and toppled by successive waves of barbarians from East and Central Asia. The Huns, the Goths, the Vandals, and other groups moved into Western Europe and were terrible not only in their behavior, but in their appearance as well. They wore fur skins, were uncouth and uncivilized, and were not Christian. Just as Europe began to recover from this onslaught, the Vikings of the North swept down, in the 900s, to scourge them again. As a result, European mythology is full of crazed, rampaging, fur-covered nightmares.

The monstrous races, as well as wild men, were as close as European scholars could get to understanding human diversity and man's primitive early state. What must be remembered is that, although legends and depictions of the monstrous races were attempts by scholars to account for and make sense of man's early development, they were not evolutionary in intent. These sensational images were popular, and the motifs of monstrous races, barbarians, and wild men were copied over and over for entire ranges of books and pamphlets. The categories of books in which monstrous races most often appeared were travelers' tales and bestiaries.

The most popular travelers' tale was that of "John Mandeville," a man who was almost certainly a literary fiction. Mandeville traveled extensively throughout the Middle East, India, and Asia, and

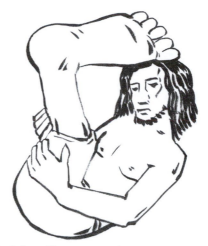

The Sciopodii *were a popular monstrous race who used their single giant foot for some interesting things.*

the likely author may have been a man known as "the Physician of Liège," though the genuine identity of the author remains under contention. In the book commonly called *Mandeville's Travels* (first published in illuminated manuscript form in the 1430s), a wide range of monstrous races appear including cannibals and the *Blemmyea.* The book was copied and recopied in several different languages and was the Renaissance equivalent of a best seller. The other popular works that often contained monstrous races were bestiaries. The first books of natural history, bestiaries, contained descriptions and illustrations of genuine animals as well as monstrous races. These works gave Europeans catalogs of the wide range of animals found outside their world as well as fantastic beasts they might encounter.

By the late 1400s, natural history collecting began. Protozoologists carefully collected and described animals and plants. Out of this work, the first accurate biological systemization grew; out-of-touch authors were being supplanted by field naturalists. With more accurate descriptions of the people of far-off places, the monstrous races began to disappear from books. Fascination with monsters did not dissipate, however, it simply shifted to human oddities. Deformed humans had always been subjects of interest and fear, but they now replaced traditional monsters. Europeans, especially the Victorians, could not seem to get enough human oddities. Fat boys and girls, dwarfs and giants, legless and armless people, and others were shown in circuses, museums, and other public attractions and even kept as pets by the wealthy. They were feted as celebrities by royalty and commoners alike. They were usually not seen as wild, and were more likely to be dressed in contemporary clothing than furs.

Initially running parallel to the image of the wild man and the caveman, but then melding with it, was the image of the ape that was becoming more and more caught up in the discussion of monstrous races and human evolution. If European intellectuals were disturbed by how closely people of color stood next to them on the ladder of life, they were even more concerned about the primates. This seemed the most disturbing relationship of all. Though ape and evolution cartoons and references in literature had been appearing in England since the 1840s, with the appearance of *The Origin of Species,* the ape and the image of the brute became a popular device for insulting individuals and entire groups. The fascination with brutes, both real and imagined, begun with the monstrous races continued. Popular culture began to sag under the growing weight of the monkey imagery of novels, newspaper articles, learned books, and cartoons that alternately ridiculed, satirized, supported, and condemned evolution and

man's bestial nature. Authors as diverse as H. Rider Haggard, Joseph Conrad, and Edgar Allan Poe all explored the interaction and relationship of monsters, men, monkeys, and society. They questioned the fine line between the civil and the savage. Robert Louis Stevenson's story of *Dr. Jekyll and Mr. Hyde* (1866) literally turned a man into a monster. The ape as the new icon of the monstrous was used to disparage ethnic groups from Africans to the Irish, to poke fun at inept politicians and brutal military leaders, and to sell products. Evolution, like monstrous races, dredged up man's relationship to God, ethnicity, class, the spread of empire, and gender issues. Where once it had been the monstrous races that filled this role, the ape and the caveman now became the metaphor for everything dark and troubling in Western minds.

By the middle of the 19th century, and with more accurate knowledge of the peoples of the world and the appearance of evolution theory, one set of monstrous races was replaced by another in the form of fossil-related monsters. The first protohumans discovered, the Neanderthals, were first pictured as fearsome creatures little different from wild men or the *Blemmyea.* Well into the 21st century, religious fundamentalist rejection of human evolution can be put down, in part, to fear of being related to ancient humans rather than Adam and Eve.

The monstrous race concept is known in all cultures, not just Anglo-European ones. Descriptions of the other in their monstrousness tell more about the culture doing the describing then those being described. A culture's fears and anxieties about its place in the world are given a face when they describe another group of which they have little or distorted knowledge. The monstrous races of the Middle

Ages can be seen now as quaint and colorful versions of the things we fear today. When our view or our minds are blocked off with ignorance, distrust, and hatred, the monsters appear.

See also: Monstrous Births.

Further Reading

Bovet, Alixe. 2002. *Monsters and grotesques in medieval manuscripts.* London: The British Library.

Friedman, John Block. 1981. *The monstrous races in medieval art and thought.* Cambridge, MA: Harvard University Press.

MOON HOAX OF 1835

The Moon Hoax of 1835 refers to a series of early-19th-century newspaper articles purporting to describe the events surrounding the discovery of life forms on the moon. Hatched by struggling British writer and *New York Sun* editor Richard Adams Locke (1800–1871), the stories ran during 1835 and claimed that real-life astronomer Sir John Herschel (1792–1871) had observed plant and animal life, including diminutive, bat-winged humanoids, cavorting on the moon. The stories were so gripping that the *Sun*'s circulation jumped to be greater than that of the *Times of London* in just a few weeks. The stories sounded so convincing that scientists from a number of countries wanted more details. Eventually, the stunt raced out of Locke's control and he began to admit it was a hoax, though mostly he claimed it was a satire. Locke's genius was to couch his descriptions in scientific-sounding language, betting his audience would not question him. This story illustrates how purveyors of pseudoscience can accomplish a

great deal to promote their ideas if they lace their work with scientific, or at least scientific-sounding, language.

Further Reading

Goodman, Matthew. 2008. *The sun and the moon: The remarkable true account of hoaxers, showmen, dueling journalists, and lunar man-bats in nineteenth-century New York.* New York: Basic Books.

Regal, Brian. 1998. When beavers roamed the moon. *Fortean Times* 109:28–30.

MOON LANDING HOAX

The claim that the Apollo moon landings of the late 1960s and early 1970s, in which humans walked on another celestial body for the first time, were faked. Believers in the hoax theory claim that it is physically impossible for humans to travel to the moon and return safely, therefore a vast and complicated conspiracy was constructed to make the staged landings seem genuine. The fake space program concept began as early as the flight of Apollo 8 in December 1968. The hoax story took off, however, with author William "Bill" Kaysing (1922–2005) and his book, *We Never Went to the Moon: America's Thirty Billion Dollar Swindle* (1981). After earning a degree in English, Kaysing went to work for Rocketdyne, the company building the engines for the Saturn V rocket, the craft that launched the astronauts to the moon. Kaysing worked as a technical writer, librarian, and head of publications; he had no training as an engineer. He worked for the company from 1956 until 1963. Not uncommon among pseudoscientists, Kaysing argued that he did not need any specialized training or education in order to see that there was a hoax or that NASA did not have the technical ability to get to the moon. In his book, originally self-published in 1974, he put forward a number of proposals that have become the bedrock contentions of hoax believers. He also claimed NASA had intentionally set the fire on Apollo 1 that killed all three astronauts aboard, had engineered the Challenger disaster, and murdered astronauts who were threatening to expose the conspiracy.

Another moon landing hoax proponent was Ralph Rene (1933–2008). Like Kaysing, Rene had no technical training, yet argued that he had figured out how the American military industrial complex had faked the moon landings. He was an autodidact with an interest in engineering. His self-published book *NASA Mooned America* (1992) covered the standards of moon hoax ideas: funny lighting, no stars in the sky, and too much radiation. Other hoax proponents have made documentaries about it and have harassed astronauts, trying to trick them into admitting they had lied. One, Bart Sibrel, accosted Apollo 11 astronaut Edwin "Buzz" Aldrin at a public event. Fearing for his and his companion's safety, the aging Aldrin punched Sibrel in the face.

Hoax proponents put forward inconsistencies, supposed scientific conundrums, theoretical problems with the technology, anomalous aspects of officially released photographs, and misleading artwork rather than any overall view of how and why the hoax was supposedly perpetrated. The cornerstone of the hoax idea—there is no central unifying theory—lies in the huge number of photographs officially released by NASA documenting the Apollo program. Hoax proponents claim that a close reading of many of these photographs reveal little glitches and clues that prove they were faked. These inconsistencies and glitches were put into the photos by a shadowy group of insiders and whistle-blowers who, it is believed, were so outraged by the hoax that they included little hints in the pictures in hopes that others would see them and expose the conspiracy. These photos show shadows that do not fall the way shadows should, and objects are lit from multiple angles when only one light source, the sun, exists on the moon. They claim the American flag put up by the astronauts waves in a breeze that should not exist, as there is no wind on the moon. Furthermore, they point to the lack of blast effect on the ground under the Lunar Module, and note that no stars can be seen in pictures that claim to show space.

There is an enormous amount of evidence supporting the fact that NASA did indeed fly humans to the moon and that they walked upon the lunar surface. It was an enormous and very public undertaking that directly involved tens of thousands of people in hundreds of companies. The public was invited to come to Florida to watch the launches, and thousands of military personal were involved in recovering the astronauts at sea upon their return. The evidence of a hoax is scanty and easily explained away. Contrary to Bill Kaysing's and others' assertions, a little technical knowledge of how the moon landings were done and how cameras work or the nature of light and shadow would help hoax proponents understand why they see the things they see. Not understanding what a phenomenon is, having knowledge of what something is, and having no experience seeing it can lead a person to misinterpret data. This is a common problem for people opposing something based solely on "common sense." The eyes can be easily fooled. One of the unfortunate aspects of pseudoscience is that proponents often have an anti-intellectual attitude. They see trained scientists or scholars as untrustworthy *because* they are educated and experienced in their specialty, and they equate training and education with lying and obfuscation. At the same time, they equate the lack of training and knowledge with honesty and truth. Moon hoax proponents also tend to be government conspiracy advocates who see themselves as heroic loners fighting against huge, entrenched, dangerous bureaucracies. In this way, they can envision themselves as true scholars seeking some grand truth that they can bring to the sheep-like masses.

See also: Hidden History; Moon Hoax of 1835.

Further Reading

Bennett, Mary D., and David S. Percy. 2001. *Dark moon: Apollo and the whistle-blowers.* Kempton, IL: Adventures Unlimited Press.

Plait, Philip C. 2002. *Bad astronomy: Misconceptions and misuses revealed, from astrology to the moon landing "hoax."* New York: Wiley.

NECRONOMICON, THE

A magical book that appears in the work of macabre short story author H. P. Lovecraft (1890–1937). *The Necronomicon* is a purported medieval text compiled by the occultist known as "The Mad Arab" Abdul Alhazred and is a compilation of the most diabolical and dangerous occult lore of the age. A number of Lovecraft's darkest characters, including the infamous 18th-century Rhode Island wizard Joseph Curwen, employed the book for nefarious ends. Lovecraft's description of the book was so vivid that some readers came to think it was a genuine

work and pestered librarians for copies. Some fans, understanding the fictional nature of the work, tried to have catalog references for it put into library collections as a joke. Attempting to cash in on the legend, several book publishers have released versions that can still be purchased. Other authors have used the book as a prop in their stories as *homages* to Lovecraft, further establishing the work as a literary touchstone. Lovecraft himself wrote a pseudohistory of the book in 1927 that was later published as *A History of the Necronomicon* (1938). He described its author's life, where it was published, and what it contained. The reality of *The Necronomicon* is that it is entirely a product of Lovecraft's imagination. An autodidact historian and antiquarian, Lovecraft regularly used actual historical personages, places, rare books, and artifacts in his writing. He also created towns, like Arkham, Massachusetts, a college, Miskatonic University, and books like *The Necronomicon* to add verisimilitude to his work. They often appeared a number of times across his work, giving it a feel of a connected set of stories.

See also: Pseudohistory.

Further Reading

Daniels, Les. 1975. *Living in fear: A history of horror in the mass media.* Cambridge, MA: Da Capo Press.

Lovecraft, H. P. 1943. *The case of Charles Dexter Ward.* Sauk City, WI: Arkahm House.

NOAH'S ARK

Biblical boat built by Noah and his family at the behest of God to save the animals of the world when He destroyed it as a

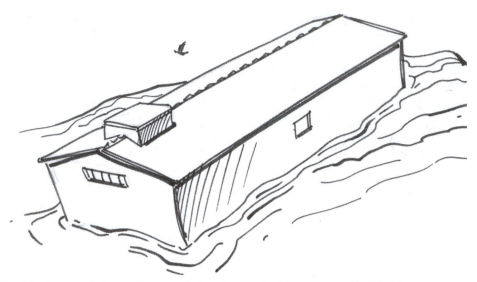

This is just one variation on the many designs for Noah's Ark put forward by Christian fundamentalists.

result of man's wickedness. The story of Noah appears in Genesis and also in the Koran. (There is also a similar story in the much-earlier Babylonian *Epic of Gilgamesh*.) When the floodwaters receded, the Ark came to rest on a mountaintop called "Ararat." From there, Noah and his family, along with the animals, fanned out across the landscape to repopulate the world. This is one of the most popular and colorful stories of the Bible and has inspired much speculation as to how Noah could have built an Ark, fit all the animals into it, where it landed, and if any traces of it could still be found. There have been numerous expeditions to find it and, while some have found tantalizing clues, most have been inconclusive.

By the Renaissance, scholars and theologians were attempting to reconstruct the Ark on paper and to determine what it looked like and how many animals it held. Anthansius Kircher (1601–1680) was troubled by the way the account compared to the reality of the growing knowledge of animal and plant distri-bution around the world. Later scholars found it even more difficult to explain how so many animals could have been transported by a single boat, no matter how big.

For most readers, the story of Noah's Ark is an entertaining fable that teaches important lessons about honor, respect for the wisdom of elders, and forthrightness. For such readers, the problematic aspects of the story pose no problem. Christian fundamentalists and Biblical literalists, however, are dependent upon every word of the Bible being true and so search for evidence that it is. Finding the Ark, they think, would support their position. As previously noted, the search for Noah's Ark has centered on Mount Ararat, in Turkey. Believers argue it is the Ararat mentioned in the biblical account. (Moslem accounts have it on *al-Judi*, a mountain on the eastern shore of the Tigris River near the city of Mosul in modern Iraq.) If found, it would constitute an extraordinary link between the pre-flood and postflood worlds. A number of Christian-believing expeditions have traveled there since the 19th century.

Supporters of the Ararat scenario look as far back as the writings of the Jewish historian Flavius Josephus (37–100) who, using yet older sources, said that people were traveling to the spot and taking away bits of wood they attributed to the Ark. Since then, there have been a number of incidents in which wood planks have been brought out and photos taken of boat-like structures. Aerial photos taken during World War II showed more boat-like images on Mount Ararat. In 1959, a better and closer photo of this same boat shape appeared in *Life* magazine, which helped fuel a popular resurgence of interest in the Ark. A search of the site in 2000 found metal plates and deteriorated rivet heads. Supporters had the artifacts chemically tested and gleefully announced they were made of modern metal (how modern metal parts would support the Ark, that, if it did exist, would be as much as four millennia old, was not explained). Researchers claimed to have made test core drillings into the site that showed traces of animal dung and cat hair under the boat shape. Satellite photos of Ararat suggest to Christian adventurer Daniel McGivern that "we're 98 percent sure" the Ark is there. That was in 2004.

See also: Anomalous Fossils; Pseudohistory.

Further Reading

Mayell, Hillary. 2004. Noah's Ark found? Turkey expedition planned for summer. *National Geographic News*, http://news.nationalgeographic.com/news/index.html.
Moen, Alan. 2007. *Noah's Ark, discovering the science of man's oldest mystery*. London: Clear View Publishing.
Woodmorappe, John. 1996. *Noah's Ark: A feasibility study*. Dallas, TX: Inst. for Creation Research.

NUMEROLOGY

Numerology investigates the relationship between numbers and events and occurrences, people's lives, and the future and is a form of divination or fortune-telling that takes groupings and patterns in numbers and reads them for significance. Each number has occult meaning, for example the number 6 (six) means a responsible character; while 7 (seven) indicates a person is fond of reading. Numbers are generated by adding the number of letters in a person's name. Birthday numbers are especially important for determining a person's character. The sequences are believed to show how a person's life will progress, how they should invest their money, marry, or make important decisions. While numerology was part of the Hebrew Kabbalah, it was also part of early mathematics, especially in the work of Pythagoras, and it is now considered pseudomathematics by modern scientists.

See also: Divination; Geomancy.

Further Reading

Ouaknin, Marc-Alain. 2004. *The mystery of numbers*. New York: Assouline.
Schimmel, Annemarie. 1994. *The mystery of numbers*. Oxford: Oxford University Press.

ORANG-PENDEK

The Orang-Pendek is a diminutive anomalous primate in the style of Sasquatch that inhabits the mountainous jungles of Sumatra. The creature, said to only be three or four feet tall, has been seen in the region for a least a century. Recent discovery of the remains of a diminutive hominid *Homo floresiensis* (commonly called the Hobbit) has led some cryptozoologists to suspect the Hobbit might account for legends of the Orang-Pendek. Found in 2003 in Liang Bua Cave on the island of Flores, the Hobbit seems to fit the common description of the Orang-Pendek and lived as recently as 12,000 years ago.

See also: Anomalous Primates; Cryptozoology.

Further Reading

Forth, Gregory. 2005. Hominods, hominoids and the science of humanity. *Anthropology Today* 21 (3): 13–17.

The Orang-Pendek is a diminutive, Sasquatch-like anomalous primate from Sumatra.

OVNI

OVNI is the Spanish translation of UFO (unidentified flying object). The late 20th century saw a dramatic increase in OVNI sightings in Latin America. The most spectacular and publicly witnessed were those over Mexico in the 1990s. OVNI photographs and film tend to be more detailed than other UFO materials. The public face of the OVNI investigation movement in Mexico is journalist Jaime Maussan. He has presented many television studies and documentaries on the subject. His most recent cause has been the promotion of a series of filmed images of an object in the sky that he calls a "flying humanoid lizard." Maussan has been hailed as a visionary chronicler of OVNI activity and has gained a large international following. He has also been criticized as being at best gullible and too easily impressed by OVNI film and photos, and at worst a self-aggrandizing hoaxer, the latter because of his involvement and support of the Dr. Reed case of the late 1990s. A man calling himself Dr. Reed, a psychiatrist, claimed to have had access to a man who had shot an alien and put it in his freezer. The case was found to be a hoax. The term OVNI is also used in French, Italian, and Portuguese.

See also: Unidentified Flying Objects (UFOs).

P–R

PALMISTRY

Also known as chiromancy or palm reading, palmistry is a divinatory art that aids the adept in reading an individual's character and future by the lines of the hand. Thought to originate with Indian astrology (the Sanskrit word is *Jyotish*) five millennia ago, it spread quickly around the world and is found in most cultures in some form. Reflexology, the rubbing of the hands for therapeutic effect, is an alternative medicine outgrowth and is meant to balance the system. There are a number of lines and raised surfaces on the human hand. This is the result of the hand evolving to be able to flex, grip, and use opposable thumbs as well as the related positioning of the musculature and tendons under the skin. These topographic features appear on all human hands, but like fingerprints, they vary from one individual to the next. These are the variation that chiromancers read to tell fortunes. Each line has a name and a corresponding meaning: the heart line, head line, life line, and fate line. The subtle positioning of these lines and raised surfaces give insight into the character of the hand's owner. As with all forms of divination, the results of palm reading are couched in vague terms that can apply to any person. Who among us *is not* worried about the future or financial security?

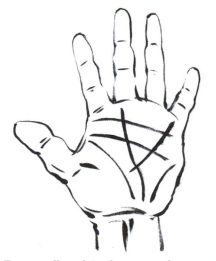

Fortune-tellers claim they can read a person's character from the lines of the palm.

125

See also: Alternative Medicine; Astrology; Divination; Geomancy.

Further Reading

Fincham, Johnny. 2005. *The spellbinding power of palmistry.* Somerset Green Magic.

West, Peter. 1998. *Complete illustrated guide to palmistry: The principles and practice of hand reading revealed.* New York: Thorsons/ Element.

PARADIGMS

A concept put forward by philosopher of science Thomas Kuhn (1922–1996) in the landmark work *The Structure of Scientific Revolutions* (1962). A paradigm is an underlying belief or set of related beliefs, that is widely accepted as the answer to a central question and around which all study in a field is based. Kuhn argued that, once established, it was difficult to overthrow a paradigm (pronounced paradime) because it had become part of accepted scientific thinking, it had evidence to support it, and careers were built around it. New ideas are forced to go up against any entrenched worldview. However, far from rejecting anomalous data in the way Charles Fort argued science did, Kuhn argued that anomalous data would build up until it reached a point of critical mass. It would then instigate a crisis in the field that would either show the anomalous data to be faulty or not compelling enough, or it would force a restructuring of the field to accommodate this new material. Accordingly, the new ideas must have great weight and a preponderance of evidence and must be ready to be shot down in its long uphill climb to acceptance. But, Kuhn argued, if the new idea that results from the anomalous data has what it takes, eventually it will be proven and accepted, it will be used to solve problems, and will eventually become the new paradigm. Classic examples of paradigm shifts—the changeover from an old to a new paradigm—are the Copernican view of a sun-centered solar system and Darwinian evolution. Supporters of these ideas had to fight against great odds to overthrow old ways of thinking.

Critics of Kuhn, among them a number of pseudoscience advocates, argue that the Kuhnian paradigm concept is misleading and unreasonable. They charge that paradigms are not created and changed based upon empirical evidence, but on intangible qualities such as the consensus of a field and other nonscientific effects. Some also argue that it is this structure that keeps new ideas and older ones, like antigravity, spontaneous human combustion, and alternative medicines, from gaining acceptance. One of the great pseudoscience advocate claims is that the scientific community or establishment is a deeply conservative structure that suppresses new ideas and focuses on propping up old and out-of-date concepts to protect jobs and positions, rather than being open to new possibilities.

See also: Falsification.

Further Reading

Kuhn, Thomas. 1996. *The structure of scientific revolutions.* Chicago: University of Chicago Press.

Maricle, Brian Andrew. 2008. *Thomas Kuhn in the light of reason.* n.p.: Light of Reason Pub.

PAST LIFE REGRESSION

The idea that, through hypnotic techniques, an individual can access memories of past events that have been forgotten or repressed or memories of lives lived before the present. Belief in past life regression (PLR) must presuppose two basic ideas: that a memory is a fixed unit that can be accessed whole and intact like a computer document, and that the metaphysical aspect of an individual—their consciousness, character, or soul—can pass from one physical body to another. A number of religions, most famously Buddhism, believe this happens. In this spiritual context, it is known as reincarnation.

In order for PLR to work, an individual must be hypnotized. Hypnosis (from the Greek *hypnos,* to sleep) is a technique for bringing a subject to a point of deep relaxation that borders on sleep. The subject can then be influenced by the hypnotist to answer questions or perform various acts. However, the subjects themselves can also use such techniques to take control of their own behavior or sensabilities. Such techniques have a long history in the human experience, mostly as ritual exercises. An early modern Western practitioner of hypnosis was the Austrian Franz Anton Mesmer (1734–1815), from which the word *mesmerism* derives. Mesmer argued for the existence of "animal magnetism," or a magnetic fluid contained in every living body. His work was not well received in scientific circles, and hypnotism came to be seen as a fraud. By the late 19th century, hypnotism had lost a good bit of its pseudoscience aspect, and it is currently accepted as a valid technique by the American Psychological Association for use with patients to overcome addiction and to help with pain management.

The more problematic aspect of hypnosis is the realm of PLR. The technique is most commonly used to recover either an individual's repressed traumatic memories or memories of past lives. Ufologists claim that part of the abduction experience involves losing time. Self-appointed alien-abduction memory-recovery specialists John Mack, David Jacobs, and Budd Hopkins all use hypnosis to help their patients remember their encounters with aliens. They claim the aliens have the power to cause abductees to be unable to remember the "missing time" they experience. Hypnosis allows for the recovery of these memories, but it can be a traumatic experience as subjects remember various tests and experiments performed upon them, experiments that are often painful and embarrassing. Alien abductees swear by the hypnotic regression technique— here called recovered memories—as one that helps fill in a painful gap in their lives.

Past life memory is just as unlikely as alien abduction memory. This is the idea that humans live multiple lives. Once one life is over, the physical body deteriorates, but the individual's life force or soul is reborn into another body. This process is often called reincarnation, which means to be made flesh again. Reincarnation is a central idea of most Eastern religions, including Hinduism, Jainism, and Buddhism. Western monotheistic religions generally reject the notion of reincarnation. Some Renaissance Christian theologians did consider reincarnation though, like Giordano Bruno who was burned at the stake by the Holy Inquisition, they did so at their peril. The late-20th-century New Age movement, however, embraced reincarnation and it entered popular culture. Soon, Western people were reporting having lived many past lives, lives far more interesting than the ones they were currently experiencing.

One such incident that gained wide attention and helped foster the spread of interest in past lives and reincarnation was the Bridey Murphy case. In 1952, a

27-year-old Colorado woman named Virginia Tighe was hypnotized and uncovered a past life as a 19th-century Irish woman named Bridey Murphy. While under the control of a neighbor and amateur hypnotist, Morey Bernstein (1920–1999), Tighe spoke with a heavy Irish brogue, sang Irish songs, and was quite entertaining in a stereotypical Irish way. Bernstein wrote up the case as *The Search for Bridey Murphy* (1956), which became a major bestseller and stirred the craze for past life regression therapy. In the book, Bernstein used the pseudonym of Ruth Simmons for Virginia Tighe. It seemed incredible that a Colorado woman with no apparent connection to Ireland could know such detail unless she had lived that life. Following publication of the book, skeptics began investigating her details, something the publisher apparently had not done. Holes and inaccuracies immediately began to appear, though supporters pointed to details that were accurate. The son of Jewish parents who had barely escaped pogroms in Russia, Morey Bernstein had become fascinated by hypnosis and was trying it on anyone he could get to do it. Along with the usual parlor tricks, Bernstein thought he might be able to go back in a person's life. He was not a psychologist and had no medical training—he was a businessman—so he had little idea of what the consequences might be, what the quality of recovered memories was, or even the nature of memory. It was later discovered that Virginia Tighe had grown up in Wisconsin across the street from an Irish woman named Bridey Murphy Corkell.

Skeptics point out the complex nature of memories and how easy it is to alter them, both consciously and unconsciously. They argue that memories are not written down in a log book in an order that can be easily accessed in their original form. Rather, we often comingle memories, construct new ones of events we did not engage in, and remember things not the way they happened but the way we wished they had happened. As such, recovered memories are highly suspect in their accuracy. When self-described alien abductees visit a hypnotherapist, they already think they have had these experiences. With the case of hypnotists like Mack, Jacobs, and Hopkins, both sides are ready to believe. This makes it almost certain that an abduction memory will be recovered.

Another problem with past life regression and reincarnation is the concept of the soul. This is the spirit element thought to transcend from one physical body to the next. But what is the soul? Most cultures have some equivalent to the Judeo-Christian soul or life force. Clear definitions of what this is, however, are hard to come by. The soul is generally thought to be an intangible quality. Christian philosopher Thomas Aquinas called the soul an "act of the body." The best we can say about the soul, from a scientific point of view, is that its definition is so vague as to render it meaningless outside the realm of the metaphysical. This would seem to place it outside the bounds of scientific quantification, let alone simple photography or recovered memory.

One of the more interesting aspects of PLR is the tendency for those who claim it to have had fascinating past lives, often as important personages. Subjects remember being kings, queens, and princes, or soldiers fighting epic battles, performing daring deeds, and being abducted by aliens. It is rare indeed for a PLR subject to recall being a dull-witted stable sweeper, sewer worker, village idiot, or other member of the social residuum. In the few cases where subjects did recall lives as commoners, they are commoners with extraordinary characters full of life, not dullards. It would seem, then, that one only gets to have a truly extraordinary past life unless the one being lived is less so. PLR can be seen as an individualized version of the golden age or hidden history theory: the idea that the past was better than the present.

See also: Alien Abduction; Metaphysics.

Further Reading

Bernstein, Morey. 1956. *The search for Bridey Murphy*. New York: DoubleDay.

Lynn, Steven Jay, and Irving I. Kirsch. 1996. Alleged alien abductions: False memories, hypnosis, and fantasy proneness. *Psychological Inquiry* 7 (2): 151–55.

PHILOSOPHER'S STONE

An alleged substance with extraordinary powers, the creation of the Philosopher's Stone was the primary goal of many operational alchemists. If produced, the stone was said to be able to transmute lead into gold. It was also reputed to be able to cure disease and prolong life. To some metallurgic-minded alchemists, the stone was a catalyst. They argued that the stone did not actually change the materials. They took this line of thought from Aristotle, who argued that all of nature, including metals, sought out the perfect form of their innate selves. This explained how coal, if left in the ground long enough, would become a diamond. The coal could not be changed from the outside, but still found its perfect inner

self on its own. Another example they cited was the way a caterpillar finds its true inner self as a butterfly. The Philosopher's Stone, they believed, simply helped lead to find its innate perfection as gold. The stone was also said to help the human body find its perfect state of health and long life, even immortality. By the 18th century, the search for the stone had been equated with the search for the elevated spiritual self. Despite over a millennium of effort, no examples of the Philosopher's Stone have ever been produced. The alchemist Nicholas Flamel was reputed to have done this, but this example goes unconfirmed.

See also: Alchemy.

Further Reading

Bailey, Michael. 2006. *Magic and superstition in Europe: A concise history from antiquity to the present*. Lanham, MD: Rowman & Littlefield Publishers.

Moran, Bruce. 2005. *Distilling knowledge: Alchemy, chemistry and the scientific revolution*. Cambridge, MA: Harvard University Press.

Thorndike, Lynn. 1958. *History of magic and experimental science*. New York: Columbia University Press.

PHRENOLOGY

The study of an individual's mental characteristics through an examination of the many shapes and bumps on his or her skull. Phrenology began its life in the early 19th century as a popular variation on the earlier scholarly work of Austrian physician Franz Joseph Gall. It became wildly popular in the United States after being promoted by the Fowler brothers, who published instruction manuals, gave special lectures, and produced a range of physical artifacts, most notably porcelain heads with lines painted on them demarcating the different areas and characters that went with them.

Franz Joseph Gall (1758–1828) believed the shape of the brain, as seen through the corresponding shape of the skull that encased it, could be used to determine aspects of a person's character. He called this technique *cranioscopy* (it was his student, Johann Spurzheim, who called it *phrenology*). This new science was the leading edge to new discoveries in neuroscience. At a time when new discoveries were flourishing, Gall was the first to write about cerebral specifications along with their specific locations. In his four volume set *The Anatomy and Physiology of the Nervous System in General and the Brain in Particular,* Gall proposed the ideas, his main works of measurements and belief in the specific functions of the brain. Bumps and contours of the head were the basis of the research that led him to believe he could discern a person's traits and behaviors just from the shape of his skull. Gall believed that the actual shape and contour of the head could determine as many as 27 different characteristics in a person's brain. This belief was based on research that indicated the brain was sectioned and that operations within it were broken up and divided for different critical skills. Each compartment was believed to control a different function or characteristic, and that the contour of one's skull would ultimately determine personality and many other traits. These separate sections of the brain were believed to be individual organs that could determine all properties of a person's character. Gall created very detailed maps showing the breakdown of each section and what it controlled.

These detailed maps led many believers to accept Gall's research and even begin to practice this new science on their own. Gall's idea of the shapes and sizes of the different organs of the brain were associated with the amount of knowledge a person acquired or lost in that particular organ. Gall believed that a person who uses a particular trait more than others would cause the particular organ with which it was associated to grow in size, while those not used as much shrunk. A pit or depression in the brain's outer skull showed a decrease in knowledge or capacity for the function that was being observed. Gall's works were revolutionary, but not well received.

In 1815, Spurzheim presented the first of his many works after separating from Gall. It was this first publication that he would later use for many other publications that broke down and/or elaborated on the original. Spurzheim was one of the men who helped the notion of phrenology to spread to America, where the new science took off. The other man who greatly contributed to phrenology's popularity was George Combe. George and his brother Andrew contributed a great deal of knowledge and expertise to the phrenology field, ad the Combe brothers were among the founding members of the Edinburgh Phrenological Society, the first of its kind in Britain. The society consisted of many men with varying degrees, but mostly lawyers and doctors. George published many writings on phrenology, the first being *Essays on Phreneology.* Two of his other contributions to the phrenology field were *Constitution of Man* and *Elements of Phrenology.* His passion for this topic helped to explore and define the discipline of phrenology.

The chief promoters of phrenology in America were the brothers Lorenzo Niles

(1811–1896) and Orson Squire Fowler (1809–1887), and it was Fowler's phrenology head, with its drawn-on lines of mental faculties that became the icon of the field. Reading of the bumps on people's heads became a popular parlor game, but was rejected by scientists who saw it as a silly perversion of the important studies of brain function and intellectual capacity. The characteristics claimed to be shown in the bumps on the head included affection, guile, cleverness, intellectual weakness, pride, and vanity, to name a few. Phrenology was also adopted by early criminologists searching for any way to help predict crime or determine guilt. It was perverted even more when it was adopted by racists as a way to prove "scientifically" the inferiority of racial degenerates.

The modern studies of brain structure, genetics, and advances in the social sciences have since discredited phrenology as pseudoscientific; the bumps on one's head have little to do with character or intelligence. While phrenology was popular, it was Gall's original ideas, though flawed, that helped pave the way for the modern study of neurology.

Further Reading

Van Whye, John. 2004. *Phrenology and the origins of Victorian scientific naturalism.* London: Ashgate.

PILTDOWN MAN

Fossil of a presumed human ancestor discovered near the village of Piltdown, Sussex, England in 1912. Hailed as an example of a missing link between apes and humans, it had characteristics of both: the skull was human, while the jaw was ape-like. In the 1950s it was discovered to be

a fake; Piltdown Man had been manufactured from a human skull and an ape's jaw. Piltdown Man is now rightly seen as an example not only of seeing what one wants to see in scientific evidence but also as a cautionary tale about evidence supporting an idea that seems too good to be true. This event argues for the careful scrutiny of facts before rash judgments are made, and has become a favorite weapon for creationists to use to counter evolution theory, especially human evolution. The incident is also employed to point out either the supposed perfidy of scientists in forging evidence to support their unsupportable positions or their hubris.

Like many paleo-anthropologists of the day, Arthur Keith was looking for something that fit his expectation of what a human ancestor should be. Australian anatomist Grafton Elliot Smith (1871–1937), then working in England and Arthur Smith Woodward (1864–1944), of the British Museum, were also looking for an early ancestor that would boost their careers. In 1912, all three men got what they were looking for in the same package.

Piltdown Man had everything an early-20th-century anthropologist and human evolutionist could want: It had a modern-design skull, an ape-like jaw, and it was found in an ancient layer of strata. It fit the missing link hypothesis nicely. At the center of the story was lawyer and amateur naturalist Charles Dawson (1864–1916). Dawson hoped to enter the rarefied world of British science as a fossil and artifact finder. In 1908, workers digging gravel on the land at Barkham Manor near Piltdown, Sussex, where Dawson had a business connection to the local landowners, found unusual rocks while repairing the road. Dawson examined them and found them interesting. More bits were found and, when comparing them, Dawson thought he had parts of a human skull. In 1911 Dawson contacted an acquaintance of his at the British Museum.

Arthur Smith Woodward was a trained geologist and curator of fish at the museum. The two men had known each other since 1891, when they first worked together to describe some fossil mammals Dawson had discovered. Woodward named them *Plagiaulax dawsoni* in honor of his friend. Dawson had a long history of fossil hunting, so, while an amateur, he had considerable knowledge of fossils, artifacts, and experience with fieldwork. Eager for recognition, Dawson threw himself into excavating the Piltdown site in 1912 and sent everything he found to Woodward. They eventually uncovered more pieces of a skull, a good portion of a jaw, and several teeth.

In November 1912, a newspaper story was published in the *Manchester Guardian* announcing a human ancestor, probably the oldest ever found, had been dug up in sleepy Piltdown, Sussex. Charles Dawson and Arthur Smith Woodward presented Piltdown Man to a meeting of the Geological Society of London on December 18, 1912. Although the actual fossil comprised relatively few pieces, a reconstruction was put together that filled in all the missing parts with plaster. The fully formed skull and jaw were a striking sight. Woodward christened the creature *Eoanthropus dawsonii,* but it was always known as Piltdown Man.

While most of the Anglo-American scientific establishment supported Piltdown Man, some opposed it. Aleš Hrdlička (1869–1943), a curator at the Smithsonian Institution in Washington, DC, was at first noncommittal about Piltdown. He was puzzled as to why Piltdown had a modern-size brain but not a modern jaw that would

allow him to talk? Hrdlička was in the minority. If an observer looked hard enough, he could see whatever he wanted. British scientists were particularly proud that this important creature was one of their own. They now had a specimen that ranked with those being discovered elsewhere in Europe, like the Neanderthals and Cro-Magnons.

Piltdown Man, however, suffered from the problem of fit. It was still unclear whether the jaw and skull went together. There was a problem of philosophical fit as well. Throughout the early part of the 20th century, other fossil humans were turning up. In addition to those found in Java, more had been discovered in China at a site outside Peking. Both Java Man and Peking Man—who were later recognized as both being *Homo erectus*—gave a different story to the course of human evolution than Piltdown did. Piltdown had been in line with earlier thinking about big brains developing firs, but Java and Peking suggested the opposite. Their development suggested that a large brain may have come relatively late in the process. As the 20th century progressed, Piltdown seemed more and more an anachronism. No more material was found after 1916.

An enormous amount of literature was produced about Piltdown Man as scientists and theologians alike tried to work it into their theories with varying levels of success. By 1953, however, little in the discussion of human origins included Piltdown Man. Such was the case that when in that year a major human origins symposium was held in England, Piltdown was not even on the official list of topics to be discussed. It was discussed informally, however, by a number of the attendees, including Kenneth Oakley, Wilfrid Le Gros Clark, and Joseph

Weiner. They discussed Piltdown's problem of fit and had many questions about how the fossils were found and why the material looked the way it did (these scientists had not been part of the generation that had been intimately involved with Piltdown, and thus were not so wedded to it). These scientists were bothered by the fact that the skull had no primate morphology. The jaw was almost totally simian; only the wear on the teeth was in the human realm. The Java and Peking fossils showed a blending of primate-like and early human morphology throughout their entire structures, further, their traits were not isolated and segregated the way Piltdown's was. The question about Piltdown's big brain was, where did it come from? Why, when all other hominids being discovered had small brains, did this creature have a large one?

Oakley, Le Gros Clark, and Weiner decided to try to find out once and for all if Piltdown Man was what Arthur Smith Woodward always claimed it was, so they went to where the original fossils were kept with the plan of examining them with the latest investigatory techniques. A series of chemical tests were performed—including the new x-ray spectrograph—which showed that the teeth and jaw had been altered by cutting and filing to show wear and to obscure anatomical details that would have given the fraud away; they had been stained with chromium as well. The jaw was that of an orangutan and the skull was an early modern human: Piltdown was a fake, built from the ground up from different parts like Frankenstein's monster.

In 1914, Charles Dawson found an artifact under a hedge at the Piltdown site that looked like an ancient tool of some kind. Dubbed the cricket bat—from the British game—this object was extraor-

dinary. It was unlike anything ever found in England. French anthropologist Abbe Bruiel, who had done extensive work on the physical culture of Neanderthals and Cro-Magnons, thought the cricket bat was the result of animals gnawing on an elephant bone, not human activity. Oakley made a copy of the cricket bat by cutting a bone with a knife and staining it. His efforts produced a replica that was identical to the original. Nothing about the Piltdown discoveries were right: not the skull, the teeth, the associated artifacts, or even the animal bones found with it. Everything must have been planted. The entire assemblage had been hoaxed.

So who did it? No one knows for sure. It could have been Charles Dawson using the "discovery" as a way up the ladder of scientific success, or as a way to embarrass those who had already made it there. Some blame Arthur Smith Woodward. The latter's motivation was advancement to the position of director of the British Museum, a post he coveted and for which he was actively campaigning. Woodward also had access to all the raw materials needed to make the forgery, while Dawson had access to the site. The 30-year relationship between Dawson and Woodward makes this scenario plausible. For years researchers have puzzled over and examined every scrap of evidence in an effort to find the hoaxer's identity.

In 1996, the contents of a storage box, originally stumbled across in the attic of the British Museum in the late 1970s, were discovered to contain skeletal materials stained in the same way as the Piltdown fossils. The box belonged to Martin Hinton, who was curator of zoology at the time of the Piltdown discovery. Woodward had turned down Hinton in 1910 for a research grant. The theory is that Hinton concocted the Piltdown forgery as revenge. He knew of Dawson's incompetence and Woodward's gullible pomposity and used their weaknesses to make them look ridiculous. Woodward was enamored of the idea of a British missing link, so Hinton led him down that trail. He even created the bogus cricket bat for Piltdown Man to play with. There are various camps that champion one suspect or another, or support their man as not being the culprit.

In his 2004 biography of Charles Dawson, Miles Russell undertook a reassessment of Dawson's private papers by cross-referencing them with the correspondence of Smith Woodward and others. This allowed him to formulate a more nuanced picture of the men's relationship. Russell's investigations led him right back to the original suspect. He argued that Dawson's desire for fame as a naturalist was the consuming passion of his life. Dawson wanted scholarly recognition from the British scientific community, wanted to be elected to the Royal Society, and maybe even win himself a knighthood, all prestigious signs of accomplishment. If Russell is correct, his explanation rules out all the other suspects with the exception of Hinton, who may have been an accomplice. It does, however, leave Arthur Smith Woodward, Arthur Keith, and the others who supported Piltdown looking foolish taken in by the cunning of a country lawyer with big ambitions.

Far from being an indictment of the evil nature of scientists, the Piltdown case reinforces the underlying idea of science as something that constantly challenges accepted wisdom and never accepts any theory as closed and final. For all the crowing creationists and anti-evolutionists like to do with Piltdown Man, it must be remembered that those same creation-

ists would never have known Piltdown was a fake unless scientists, who were questioning their own field, showed them it was.

See also: Anomalous Fossils; Creation Science.

Further Reading

Regal, Brian. 2007. Piltdown and the almost men, in *Icons of Evolution,* Vol. 2. Westport, CT: Greenwood.

Russell, Miles. 2003. *Piltdown man: The secret life of Charles Dawson and the world's greatest archaeological hoax.* Stoud, UK: Tempus Publishing.

PROTOCOLS OF THE ELDERS OF ZION, THE

Infamous forgery and faked manifesto purporting to be the plans of a shadowy group of Jewish intellectuals with ties to Freemasonry, detailing how they plan to takeover the world's banking and business apparatus. Known by several different title variations, *The Protocols of the Elders of Zion* is a step-by-step guide for world domination that has spawned numerous conspiracy theories and is still believed, despite its complete debunking as a hoax, by racist, anti-Semitic, and conspiracy theory aficionados worldwide to the present day. Meant to be a secret document for insiders only, it was supposedly leaked to a journalist for publication.

The *Protocols* had a strange genesis and history, having been modified by numerous authors and editors, and having been used for different ends by different groups. It was not written as a single work, but cobbled together from several different sources of preexisting texts—not all of them anti-Semitic—by Russian security personnel between 1895 and 1902. The Tsar of Russia, Alexander II, had been working to bring social and political reforms to the country, but hardliners within the aristocracy who were fearful of losing their positions and power conspired to have a journalist, Matvei Golovinski (1865–1920), put the book together for propaganda purposes. Making Jews the focus of the threat played into the preexisting anti-Semitism prevalent in Russia and throughout most of Europe. The forgery was meant to scare the next Tsar, the future Nicholas II, who also had anti-Semitic tendencies, into backing away from reforms.

The *Protocols* were an amalgam of several texts, including *The Dialogue in Hell between Machiavelli and Montesquieu* (1864), a political parody and commentary on the French government of Napoleon III, and the anti-Jesuit novel *Mysteries of the People* from the same period. Golovinski grafted these works onto an anti-Semitic German book, *Barritz* (1868), replaced the Jesuits and Napoleon with Jews and Masons, and *The Protocols of the Elders of Zion* was born. It was then used to convince the Tsar that continuing social reforms in Russia would give into the conspiracy. In modern times it would have been said that if he did not, he was "playing into the hands of terrorists." The *Protocols* did not remain static. They were used again in the 1917 revolution against the Bolsheviks, thus connecting for the first time Communism and Jews in popular culture. Translated into German in 1919, it also helped to incite anti-Semitism there and was later used by the Nazis to justify race laws and the Holocaust of the 1930s and 1940s.

That the document was already being called a forgery in the early 1920s did not stop its influence. It appeared in its first English language form in 1919 when it was excerpted in the Philadelphia *Public Ledger,* though the Jewish references were replaced by ones concerning socialism. The next year, the racist, pro-Nazi American automobile maker Henry Ford published the *Protocols* in his *Dearborn Independent,* though not before the Jewish material was put back in. The *Protocols* have been reprinted many times in many languages. The text has been used by journalists and political commentators around the world to disparage not just the Jewish people and religion, but immigrants, foreigners of all kinds, and almost anything a given commentator wanted. They have been adopted as genuine by a wide range of groups, including neo-Nazis, the Ku Klux Klan, conspiracy theorists, antigovernment rabble rousers, religious fundamentalist Armageddonists, and Moslem extremists.

The Protocols of the Elders of Zion is an example of pseudohistory. It is a lie from beginning to end, but fills a need for bigots looking for justification to support their belief in mass conspiracies and shadowy groups engaging in mayhem. With a work like the *Protocols,* any group's name could be substituted for the Jews as the evildoers. The longevity of the *Protocols* comes not from its witty prose or its deep insights into the human condition—it is haphazardly and clumsily written, tedious, and dull—but from its ability to feed the fears, anxieties, and irrationalities of those easily fooled.

See also: Pseudohistory.

Further Reading

Ben-Itto, Hadassa. 2005. *The lie that wouldn't die: The protocols of the elders of Zion.* Edgware, UK: Mitchell Vallentine & Company.

De Michelis, Cesare G. trans. Richard Newhouse. 2004. *The non-existent manuscript: A study of the protocols of the sages of Zion.* Lincoln: University of Nebraska Press

Wolf, Lucien. 1921. *The myth of the Jewish menace in world affairs or, the truth about the forged protocols of the elders of Zion*. New York: Macmillan.

PSEUDOHISTORY

Similar to pseudoscience only dealing with historical topics, pseudohistory uses the trappings of scholarly historical method—footnotes, citations, bibliographies, and primary sources—yet is designed not as a thoughtful analysis intent on discovering insights into the past, but as a subjective arranging of facts to support a predetermined idea that often supports a political agenda. It is an accepted aspect of postmodernist historical interpretation that an historian's personal worldview affects his work. Even the most scrupulously objective researcher will still see his worldview impact, to some degree, how he analyzes material. This is an accepted element in the historical community and is easily dealt with by accepting that there is no final answer to larger questions of analysis. Multiple points of view are integral to getting as wide an understanding of historical events as possible. It is a natural and subconscious part of scholarship. Where it becomes problematic is when it occurs unnaturally and consciously for some specific end.

A simple definition of history is that it is the written discussion and analysis of past events. The past is everything that happened before the present. The past, however, is made up of many disconnected parts floating in space with no apparent meaning. Only when some of those past events are arranged, collated, and analyzed does history appear. The book you are currently reading is an historical analysis of the past; it is not the past itself. It is also not the only possible interpretation. You, the reader, after seeing the facts I have laid out might add others not covered here and thus disagree with my conclusions. That is acceptable; that is the entire point of the exercise, to get you thinking.

Like scientists, historians love to argue over interpretations of data. These discourses tend to center around analyses of ambiguously understood events rather than factual content. Not all historical events are known to the same level of certainty. There are many cases where original materials have been lost or destroyed and cases where no records of an event exist. In cases like these, such ambiguity allows for multiple interpretations. This kind of arguing is, in fact, the meat of what historians do. However, the range of interpretation is still limited by what evidence is available. For example, a politician's action at a given moment might suggest she intended to pass a law, while it might also suggest she intended to block that law. One thing that would be completely pseudohistorical would be to argue that those actions indicate she was under the control of space aliens. It would be pure conjecture supported by no evidence at all.

There are historical events where no preponderance of evidence supports that it occurred. Historians argue over the meaning of wars, plagues, and peace movements, or trends in pop culture (and every other aspect of human history) all the time. Only when the evidence is dubious do they argue over whether such events actually happened when the evidence clearly shows they did. To be seen as genuine historical scholarship, there must be evidence to support it.

An example of pseudohistory is Holocaust denial. This is the attempt to prove that the Nazi extermination program of Jews and others during World War II did not happen. Practitioners of Holocaust denial argue that what were purported to be death camps run by the German government were only internment and work camps. They acknowledge that many died in these institutions, but that the camps were not part of a systematic, state-run genocide machine, and the commonly accepted number of deaths (over six million) is a gross exaggeration. They also sometimes argue that, while cruelties did take place inside the camps, mistreatment was carried out by low-level functionaries and was not known to be happening by the Nazi leadership, including Adolf Hitler. Finally, Holocaust deniers claim that the "myth" of the Final Solution (as the Nazis called it in their coded language) was an invention by Jews to gain worldwide sympathy for their plight.

The revision of history is common. Historians learn new things and discover new facts about the past on a regular basis. This new information can sometimes profoundly alter how past events are perceived. Holocaust deniers claim they are merely revising history. However, the evidence they use is dubious and ambiguous at best. Legitimate evidence has contributed to a revision of Holocaust studies; it has never refuted or negated the basic facts the way deniers try to. The denial of the Holocaust began even before the Final Solution itself began. The upper echelon leadership of the Nazi party couched their orders in ambiguous terms and code words so that only insiders might know what those terms meant. This was to establish plausible denial of their actions and to shield what they were doing: It was a cynical act to deny hor-

rific and deplorable actions before they happened. Individuals like Hitler and his henchman Heinrich Himmler never wrote to each other saying, "Yes, now we will commence murdering the Jews," but the inferences were there. There was also a lower level bureaucracy needed to make the system work. These operational levels kept records of what they were doing, but as the war was spiraling down and the Allies began closing in, Himmler ordered his underlings to destroy the evidence linking him to the death camp system. (It is common for corrupt politicians and corporate leaders to destroy their records when they think their reign is coming to an end.) Fortunately, vast reams of material survived this attempted purge by the Nazis of the evidence for the Holocaust.

Postwar Holocaust denial began with the work of Columbia University historian Henry Elmer Barnes (1889–1968). An established academic, Barnes was a liberal critic of the New Deal and the American entry into World War II. Following the war, he began to argue that the war was the result of British and American provocation, rather than militaristic empire-building on the part of the Germans, Italians, and Japanese. For Barnes, the Japanese attack on Pearl Harbor was the American administration's fault. As for the Holocaust, instead of saying it had not happened—an idea even Barnes knew was ludicrous—he took a nuanced but even more insulting approach, arguing that German Jews shouldered as much responsibility for the Holocaust as non-Jewish Germans. A small coterie of other pro-German American historians followed in Barnes's footsteps, most notably David Hoggan (1923–1988), who is the author of what is considered the first genuine work of Holocaust denial, *The*

Myth of the Six Million (1969). He argued, among other things, that Jews didn't really have it that bad under the Nazis' infamous race laws of the 1930s. It was later discovered that Hoggan was a member of a neo-Nazi organization and had been funded by such groups. Hoggan's work was roundly criticized by other scholars in the United States and Germany, who pointed out the pseudohistorical aspects of his work. Hoggan's doctoral advisors at Harvard University also renounced him, saying his doctoral work was different from his published work. His writing was embraced, however, by neo-Nazi and anti-Semitic organizations as academic support for their views.

The best and most incontrovertible evidence supporting the reality of the Holocaust, and putting a lie to denial, comes from the Nazis themselves. In order to carry out their plans to exterminate people on an industrial scale, architects had to design the machinery of murder, contractors had to build camps and crematoria, trains had to carry victims from staging areas to the camps, troops had to be fed, and offices staffed. All this activity had to be coordinated, paid for, and otherwise recorded. Despite their efforts to destroy it, a good portion of this paper trail survived the war. The camps themselves survived physically as well. Survivors of the camps told their tales. Some of the German soldiers, their consciences wracked by the horror of what they were doing, wrote confessional letters home. When Allied troops liberated the camps, they recorded what they found. Perhaps most damning of all, the Nazis who staffed these facilities and ran the extermination machinery took photos and films of what they were doing. As one historian described it, they took family snapshots

as if they were tourists on a vacation in Hell. This preponderance of evidence all supports the fact that the Holocaust did occur.

There have been a number of Holocaust denial trials. A 1985 case in Canada saw publisher Ernst Zündel convicted and jailed for his Holocaust denying. In 1992, his conviction was overturned and the statute he was convicted upon thrown out as unconstitutional under Canadian law on the grounds that it violated the freedom of speech. Anti-Holocaust denial activists have wrestled with how best to address this issue. Wanting to avoid restricting the freedom of speech, many advocate for allowing Holocaust deniers to spread their ideas, but then showing them up for the fools they are. The most famous case is that of British author David Irving versus the American Deborah Lipstadt in 1998. In her book *Denying the Holocaust* (1994), Lipstadt accused Irving of intentionally misrepresenting facts and evidence in his work. Irving sued her in a British court for libel. Lipstadt's defense team studied Irving's work and found he had, indeed, fabricated key historical documents. Lipstadt won the case and Irving was labeled a Holocaust denier and neo-Nazi polemicist in court. When Richard Evans, the Cambridge University historian who fact-checked Irving's book for the Lipstadt defense team, published his account of the trial in 2002, Irving threatened to sue him as well. In 2006, Irving pleaded guilty to denying the Holocaust in Austria, where it is a crime to do so. Holocaust denial has spread from the United States and United Kingdom to France and into the Middle East, where it has been adopted by religious fundamentalists and anti-Israel proponents. In a 2005 speech, the President of Iran, Mahmoud Ahmadinejad, famously called the Holocaust a fairytale designed to gain sympathy for Jews and the State of Israel. There have been similar dismissals of genocide by the Turkish government of the Armenian genocide during World War I and dismissals of genocide concerning Rwanda and Bosnia, both in the 1990s. Holocaust denial can, in part, be explained by political bias and darker motivations. The greater meaning of the Holocaust can be and is argued; that the Nazis purposefully and willfully exterminated over six million people based solely on their religious persuasion cannot. This makes Holocaust denial an example of pseudohistory.

Pseudohistory in its less nefarious form can be used to make a literary work seem to have a place in time and sense of reality. This is where an author places genuine historical figures, places, or events into a fictional story. A good example is Caleb Carr's *The Alienist* (1994), in which Carr has real-life Commissioner of New York Police Department Theodore Roosevelt assist a team of investigators to catch a serial killer. Carr, however, did not put forward his story as a chronicle of real life events. When Dan Brown wrote *The Da Vinci Code* (2003), he ambiguously suggested in the front piece to the text that, while the actual story was fictional, the details of the operations of the Catholic Church, Opus Dei, and the Priory of Zion were accurate. Many readers, unaware of the fictional nature of the story, took it at face value. Historians and church officials alike were not pleased.

The promotion of pseudohistory is more than an intellectual exercise. It is often used as a political tool to cover up past horrors of which the perpetrators are ashamed, to make politicians seem to have been more successful than they really were, or, in the case of golden age

and hidden history theories, to make the current state of the world seem far worse than it is. Pseudohistory is often a component of conspiracy theory as well. We have to guard against pseudohistory, lest the reality of the past be swept away and lies and falsehoods replace it.

Further Reading

Daston, Lorraine, and Peter Galison. 2007. *Objectivity*. Cambridge, MA: Zone Books.

Evans, Richard J. 2002. *Lying about Hitler: History, holocaust, and the David Irving trial*. New York: Basic Books.

Lipstadt, Deborah. 1994. *Denying the Holocaust: The growing assault on truth and memory*. New York: Plume.

Novick, Peter. 1987. *That noble dream: The 'objectivity question' and the American historical profession*. Cambridge, NY: Cambridge University Press.

RED MERCURY

Supposed substance used in the manufacture of nuclear weapons. Described in a number of ways—which do not all coincide—the properties of red mercury render it a kind of modern-day Philosopher's Stone. It is capable of a wide range of services and applications. It is simultaneously toxic and curative, and is also said to be a crucial component of atom bombs, but nuclear weaponeers are at a loss to determine what it does. However, British/American physicist and influential bomb designer Samuel Cohn claimed red mercury was a cover name for a chemical explosive he called *Ballotechnic,* and which he said could be used to build everyone's nightmare, the suitcase nuclear bomb. Cohen's specialty was nuclear bomb radiation and death rates. During the Vietnam War, he famously suggested the U.S. government use small nuclear devices against the North Vietnamese and Viet Cong. There are no confirmed examples of this substance, which only seemed to enter media reports in the mid-1980s as the Cold War was reaching its crescendo. It is commonly thought now that red mercury was a sting operation set up by the U.S. government to lure nuclear black marketeers into the open for capture.

Further Reading

Commander X. 2002. *Red mercury: Deadly new terrorist super weapon.* New Brunswick, NJ: Inner Light Publications.

ROSWELL INCIDENT

Legend of the crash of an alien spacecraft in the New Mexico desert near the town of Roswell in 1947, the Roswell Incident is the most enduring and contested event in all of unidentified flying object (UFO) lore. The unusual wreckage was initially discovered by a rancher, but shortly after personnel from the nearby Air Force base arrived and recovered the materials. Local people who knew of the incident or who had seen the wreckage were told by government operatives not to discuss what had happened. Initially, the air base announced publicly that a "flying disk" had been captured. The sensation generated by such an announcement was countered within a few days when the air base authorities announced it was not a flying disk after all, but a crashed weather balloon. The excitement soon diminished and the event was forgotten until it was resurrected, in altered form, in the late 1970s and 1980s.

The Roswell Incident is the most discussed, researched, and problematic of all UFO stories. That a UFO crashed in

the New Mexico desert is supported by no physical evidence, only the murky and changing testimony of numerous eyewitnesses. Over the years almost all the early reports have been embellished and altered, and crucial aspects of the story—like the recovery of both dead and living crewmembers of the craft—did not appear until years after the fact.

All that is known for sure about the case is that something did crash land on rancher Mac Brazel's property in August 1947. A collection of military personnel did retrieve these materials, and the air base public relations office did announce it had captured a flying disk, only to change the story later to a weather balloon. The story was resurrected in the late 1970s when base intelligence officer Major Jesse Marcel told a ufologist that the materials he helped collect at Roswell were very unusual and that he thought they were not of this earth. His story was supported by the testimony of his son, Jesse Jr. who remembered his father bringing home bits of the material for him to see and that his father claimed the pieces were from a spacecraft. With the dissemination of this material, other Roswell inhabitants began to tell their stories. It was at this point that revelations about sinister military personnel and alien bodies emerged. Speculation grew as to what had happened, and the once-forgotten Roswell Incident took on cosmic proportions of interstellar contact and nefarious government cover-ups.

Believers argue that an alien ship crash landed in the Roswell area while observing military operations at the local air base (where what was at the time the only nuclear bomb-armed air unit was stationed) and at the nearby White Sands test range. A number of alien crewmen were killed in the crash and survivors were whisked off to disappear in the

U.S. military/industrial complex. Military personnel arrived on the scene *en masse* and threatened the local populace into silence, lest they disappear as well. A vast conspiracy was then set into motion to hide these facts. There are enormous inconsistencies and story gaps in this explanation to go along with the dearth of physical facts. One example is that Jesse Marcel claimed that the pieces of the ship he handled were made from a material that defied description: It was flexible enough to crumple into a ball, yet could be neither cut with a knife nor burned with a flame. It was essentially indestructible. However, the crash site was described as being littered with pieces of the broken ship. So, this incredibly advanced material that can neither be cut, torn, nor burned and which returned to its original shape when crumpled and could hold up to the rigors of interstellar flight shattered into 1,000 pieces upon a rough landing in the desert.

In the 1990s, it was discovered that a top secret military project called Operation Mogul was being conducted at the time at the White Sands Missile Range. This consisted of elaborate balloons linked to sensitive recording equipment meant to search for traces of Soviet nuclear tests in the upper atmosphere. It was one of the most secret projects in the U.S. military at the time, rating the same level of secrecy as the Manhattan Project. At the time Mac Brazel found debris on his land, a Mogul balloon had malfunctioned and crashed in the area. The U.S. Air Force sponsored a pair of books, *Roswell Report: Fact or Fiction* (1995) and *Roswell Report: Case Closed* (1997), that laid out details of Operation Mogul and two later projects, codenamed High Dive and Excelsior. These last two were tests to see how far above the earth an air crewman could parachute and still be likely to survive. High Dive used anthropomorphic dummies, while Excelsior used actual men. It is the military's contention that the Roswell Incident was the result of civilians coming into contact with secret experiments, misunderstanding what they saw, and confabulating events that did not take place at the same time. All of the stories about the Roswell case are consistent with military projects of the descriptions of Mogul, High Dive, and Excelsior.

Critics of the government argued that these stories of secret operations were a blind cover-up of what really happened. Numerous Freedom of Information Act requests to the government, including one from a U.S. senator to have all documents pertaining to the case be made public, have yielded little information. The government argues it cannot release what it does not have. There are inconsistencies within the military's explanation, but this story holds up better and is far more plausible. The general distrust of the U.S. government by many of its citizens, a distrust history shows is unfortunately well-earned, will keep the mystery and interest in the Roswell Incident high. If there was a crash of an alien spacecraft in the desert of New Mexico in 1947, then the Roswell case is an example of the duplicity and paranoia of governmental power. If there was no spaceship crash, it is a good example of the difficulty of proving to a believer that there is nothing there to believe.

See also: Alien Abduction; Alien Autopsy; Unidentified Flying Objects (UFOs).

Further Reading

Berlitz, Charles, and William L. Moore. 1980. *The Roswell incident: The classic study of a UFO contact.* New York: G.P. Putnam's Sons.

McAndrew, James. 1997. *The Roswell report: Case closed.* New York: Barnes & Noble.

S–Z

SASQUATCH. *SEE* BIGFOOT

SATANIC ABUSE AND CRIME

Accusations that Satanists, worshippers of the Devil, have been systematically kidnapping men, women, and children for use in their rituals and engaging in criminal acts. The abuse comes in the form of physical beatings, brainwashing, sexual torture, imprisonment, and ritual murder. Reports and accusations of such activity come from around the world. In North America, the United Kingdom, and Europe satanic-associated crime is most often perpetrated by disaffected youth. Small groups, usually with a male leader, engage in rituals that occasionally escalate into crime, violence, and, less often, murder. These cases are the most frequently documented examples of satanic crime. Perpetrators claim they are working in the name of Satan, but they usually have little understanding of the history of satanic belief; more often than not they take their ritualistic cues from pop culture, television, and film, not historical or theological sources. Rather than showing a genuine diabolic influence, the cases show sad, pathetic teenagers and young adult outcasts who are easily manipulated into antisocial and occasionally criminal behavior. However, despite the media frenzy and the pontification of church leaders, the percentage of those who embrace the satanic lifestyle is low, and those who go on to commit crime even lower.

Large-scale, group satanic crime and abuse is hard to document. Christian belief in satanic cults goes back to its origins in the first and second centuries. The modern myth of organized cult activity is often traced back to the publication of *Michelle Remembers* (1980), by Canadians Michelle Smith and her doctor, Lawrence Pazder. Smith claimed she was the victim of satanic abuse at the hands of her own mother who was part of an organized, worldwide cult called the Church of Satan. As a child in 1954 and 1955 she was forced to participate in and witness unspeakable horrors performed by cult members. Only her Christian faith, at five years old, saved her. Under hypnosis, Smith made outrageous claims about the extent and operations of the group. The book helped launch belief in the most nefarious cases of satanic abuse: those involving the victimization of children. For his part, Pazder developed a new career consulting with churches and police agencies on the kind of satanic abuse his book helped bring to light. He also became a television consultant on the topic. Pazder and Smith eventually divorced their spouses and married each other.

The Smith case echoes the 19th-century Runaway Nun cases, the most famous of which concerned Maria Monk. These stories alleged that innocent young women were tricked into joining convents and, once there, were subjected to various tortures and depravities not unlike those described in *Michelle Remembers*. In these stories, which were best sellers, Catholic priests rather than Satanists occupied the role of the narrative's evil scoundrels. These stories helped whip up anti-Catholic sentiment and generated resentment and bigotry toward Catholics that occasionally turned into physical violence.

Almost immediately after *Michelle Remembers* was released, investigative journalists dug into the case and were able to show that the allegations of the book were false, that there was no evidence to support any of it, and that it was all a work of fabrication. Subsequent investigations also showed no substance to any of the charges. Anton LeVay, founder of the California Church of Satan, sued for libel to have the connections made in the book to him and his organization removed. He won and was adjudicated an award for damages. The *Michelle Remem-*

bers controversy ignited belief in an imminent threat from demonic cults and helped bring on the most infamous case of satanic abuse, the McMartin Preschool case.

In 1983, allegations were made that the staff of the family-run McMartin Preschool in Manhattan Beach, California, had been systematically abusing their charges. One of the students' mothers, Judy Johnson, accused her husband and a McMartin school worker of sexually assaulting her child. Police sent out a flyer to the parents of other McMartin school children and a panic ensued. Hundreds of children were interviewed and a list of horrendous and simply bizarre allegations about Satanism (including teachers who could fly) were made. Arrests commenced and the school closed. Lawrence Pazder and Michelle Smith became involved as consultants to the families.

Later, it was determined that Judy Johnson, the original accuser, had mental health and alcohol abuse problems. Also, zealous local police and prosecutors had withheld vital evidence from the defense that would have shown the allegations to be spurious and vindictive. Eventually, a new district attorney dropped charges against all but two of the many defendants for lack of evidence, but did continue to prosecute the last two. These trials lasted for more than six years—the longest in American legal history—and cost taxpayers millions of dollars. By 1990, all charges had been dropped against the remaining two defendants, though not before one had served five years in jail and seen his career ruined. Accusations in the McMartin case can be put down to an angry, delusional, alcoholic mother, vindictive police and prosecutors, and an ambitious, ratings-obsessed reporter and media producer rather than ferocious devil worshippers.

In response to the McMartin case, similar accusations sprung up in England in Cheshire, Nottingham, and Rochdale. They followed the McMartin model of hysterical claims, interviews of children, law enforcement members overzealously believing everything they heard, and media working everyone into a frenzy. Allegations soon emerged in other locations across England. Like the McMartin case, no evidence other than an apparent need to believe that children do not lie, accompanied the wild claims. Charges of satanic abuse are usually brought by two categories of accusers: adults who claim they were molested as children, as in Michelle Smith's case, and when an adult convinces a child that he or she was molested by someone else. The memories of the victims are often of the "recovered" type similar to those elicited from alien abduction victims and remembered through hypnosis. When the accuser is a child, therapists and others can often take what the child says out of context or distort it, and the child will often say whatever the adult wants in order to gain favor. Accusations can reach wild heights, including the systematic rape of dozens, if not hundreds, of children by one small group of adults, along with the murders of those children. The abuse is also often claimed to have occurred in close proximity to unsuspecting parents.

Another common model is the family of satanic abusers who hand down their ways from generation to generation. Unfortunately, there is little or no corroborative evidence for this phenomena. A 1992 FBI report of an investigation into satanic abuse found no evidence of widespread organized abuse of the kind claimed by those accusers who believe it is happening. The report also failed to define ritual abuse because of its vagueness. The FBI did not claim that child abuse does not occur, only that the stereotype of large-scale "satanic-occult" abuse and organized crime was undocumented. Despite this finding, law enforcement is generally divided on the issue with some members of the police community believing claims wholeheartedly, while others take a more skeptical view. A British government report issued around the same time came to similar conclusions.

Before these cases, there were also the Charles Manson and Son of Sam cases. Both were later seen to have had satanic connections. Investigative reporter Maury Terry outlined the widespread satanic conspiracy he felt was behind two of the most infamous murder cases in American history. In *The Ultimate Evil* (1987), Terry charges that a satanic underground had prompted the Manson Family's actions and was also responsible for the murder spree of young adults that panicked New York City in 1977.

Like the McMartin case, there were law enforcement officials who thought, and still think, a group was responsible for the Son of Sam killings. Unlike the McMartin case, however, the principle character in the saga agreed. David Berkowitz, captured in 1977 and who plead guilty to the Son of Sam murders, claimed he was involved in a satanic cult that orchestrated the killings. Inconsistencies in sightings and events and multiple witnesses who gave descriptions of the gunmen that did not all look like Berkowitz convinced Terry and others that a group was behind the killings. Terry claimed a nationwide organization existed in America with connections outside the country and that this organization was responsible for numerous crimes. Public awareness of the supposed occult connections to the Mason and Son of Sam murders did not spread until the late 1980s.

See also: Alien Abduction.

Further Reading

Allen, Deena, and Janet Midwinter. 1990. Michelle remembers: The debunking of a myth, *The Mail on Sunday* (September 30): 41.

Eberle, Paul, and Shirley Eberle. 1993. *The abuse of innocence: The McMartin Preschool trial.* New York: Prometheus Books.

Frankfurter, David. 2006. *Evil incarnate: Rumors of demonic conspiracy and ritual abuse in history.* Princeton, NJ: Princeton University Press.

La Fontaine, Jean. 1998. *Speak of the Devil: Tales of satanic abuse in contemporary England.* Cambridge, NY: Cambridge University Press.

SEA MONSTERS

Like other members of the monster family, sea monsters have a Zen-like ability to both exist and not exist. While they are cryptids they are also monsters, but once they are discovered they become just another species. Aquatic monsters come in two forms, freshwater lake monsters and marine sea monsters. Unlike individual lake monsters, which tend to appear on a fairly regular basis, sea monsters are often single sightings or are found as beached carcasses. The sea has, however, produced its fair share of monsters turned accepted species. The mola or giant sunfish, the sturgeon, and various eels have all been mistaken for monsters. The best example of the phenomena is the monster called the Kraken. Reported by sailors in the waters around Norway and Iceland for centuries, the Kraken was finally identified as the giant squid.

Given that sea monsters appear in the lore of all seafaring peoples, it is safe to assume that the earliest accounts of sea monsters were undoubtedly the results of mariners encountering creatures for the first time that are commonly known today, such as whales, dolphins, and sharks. All these creatures, and others, live close to the ocean's surface and engage in behaviors readily observable from a ship. Snake-like marine animals are the origin of the term "sea serpent," which eventually became synonymous with sea monster. The work of folklorist Adrienne Mayor, focusing on Greek interpretations of fossils, suggests that the discovery of the remains of ancient marine life, such as ichthyosaurs and plesiosaurs, prior to the 19th century understanding of these creatures could have been the basis of sea monster legends. The oldest example of naturalists attempting to categorize sea monsters is the work of Swiss author Conrad Gessner (1516–1565). In the early 19th century, naturalist Constantine Rafinesque named the Massachusetts sea serpent *Megophius*. This was later modified into *Megophius monstrosus* by the late-19th-century Dutch zoologist A. C. Oudemans (1858–1943) in *The Great Sea Serpent* (1892). The 20th-century's great sea monster chronicler was Bernard Heuvelmans (1916–2001), whose work in the field helped inaugurate the modern pursuit known as cryptozoology. His *On the Track of Unknown Animals* (1958) and *In the Wake of the Sea Serpents* (1968) helped keep the search for these creatures going. The most common sea monsters to be examined close up are the supposed remains of dead individuals that appear on beaches as little more than large piles of blubber-like material, known as globsters.

The term was coined in 1962 by monster hunter Ivan Sanderson to describe such a find from Tasmania. Analyses of these globs has either been inconclusive or shown them to be remains of known marine animals, usually dead whales.

See also: Cryptozoology.

Further Reading

Carr, S. M., et al. 2002. How to tell a sea monster: Molecular discrimination of large marine animals of the North Atlantic. *Biological Bulletin* 202:1–5.

Coleman, Loren, et al. 2003. *Field guide to lake monsters, sea serpents, and other mystery denizens of the deep.* New York: Tarcher.

O'Neill, J. P. 2003. *The great New England sea serpent: An account of unknown creatures sighted by many respectable persons between 1638 and the present day.* New York: Paraview.

Sweeney, James B. 1972. *A pictorial history of sea monsters and other dangerous marine life.* New York: Crown.

SECOND SIGHT

A form of foresight in which one sees the future while in a trance, a dream, or even during a waking experience. Second sight is closely aligned with prophecy; all prophets have had the gift of second sight to one degree or another. It is also related to extrasensory perception, telepathy, and premonitions, feelings about what will happen rather than images of the future. Visions can run the gamut from crystal clear to blurry, indistinct suggestions of future events. Second sight has been reported back to antiquity, was often performed by oracles and other spirit mediums, and has been used by members of all cultures. Accounts of second sight in literature are typically associated with characters defined as mystics. Second sight is usually associated with impending tragedy, like in the scene from *Julius Caesar* in which the scrawny old soothsayer pushes his way through the crowd to deliver the best-known line of prophecy, "Beware the ides of March!" The modern variation on second sight is remote viewing. The problem with second sight, like with prophecy in general, is its poor track record of success. While it adds a dramatic touch to literature, second sight is known only from anecdotal evidence given after the fact. When academics, scholars, or political or stock market analysts make predictions, they are not calling on paranormal abilities, but on a thorough knowledge of the workings of their field. They also understand that they still can be wrong, and often are.

SHAPE SHIFTING

Shape shifting is the ability to change one's body from one form to another or to change someone else from one form to another against his will. This transformation is effected through the application of occult techniques. The most common example of shape shifting can be seen in werewolves, who are afflicted with a condition known as lycanthropy. There are also ethereal forms of shape shifting that are spiritual rather than physical in nature.

A number of phenomena are associated with shape shifting. The most well known is the metaphorical type common to literature from around the world and dating to antiquity; it is likely that shape shifting was a part of preliterate oral traditions as well. Metaphorical shape shifting is a literary trope employed to suggest both positive and negative change, to conceal something, or to trick a character into doing something he might not normally do. Literature is full of stories of gods, pixies, fairies, witches, demons, and other supernatural entities changing mortals into animals and inanimate objects. Using this technique, frogs become princes and people can be transformed into horses, dragons, or trees (often as punishment).

Another form of shape shifting is philosophical. There is a long tradition of religious adepts throughout history of engaging in meditations and profound bouts of prayer during which a euphoric state is achieved and the practitioner mentally transforms by reaching an altered state of consciousness. Sufferers of multiple personality disorders (MPD) are reputably able to change their character and become "someone else." This is also known as dissociative identity disorder. This behavior is contentious and not widely accepted in the psychiatric community, as it is relatively simple to fake. Occasionally, physical behavior changes—stooping, head ticks, and altered voices—accompany MPD, but these changes are only superficial.

The examples noted here should not be considered pseudoscience, as they are literary, spiritual, and psychological, not literal. Some scientists argue that believing MPD is a genuine condition is a form of pseudoscience, as the medical community has accepted little evidence of its existence.

The final form of shape shifting is much like the literary form noted above, only it is believed to actually occur, thus it is the most problematic from a scientific point of view. There are many examples in the biological world of living organisms that can change color or shape as protection from predators or in aid of locomotion. Some, like the puffer fish, can momentarily alter their appearance, but they remain the same creature. There are no examples in the biological world of organisms that can completely change their fundamental morphology, such as a pig transforming into a tree or a human into a shower of gold. While legends abound of such behavior among humans, there is no evidence to support such abilities.

The most recent accusation of shape shifting on a mass scale comes from British conspiracist/author David Icke. In *The Biggest Secret* (1999), Icke claims that alien, reptile-like creatures pose, through shape shifting, as most of the world's political and religious figures—most notably the British royal family—so that they might secretly rule the earth's human population for various nefarious purposes. Icke has only circumstantial evidence to support these claims, yet still garners an enthusiastic audience.

See also: Werewolves.

Further Reading

Icke, David. 1999. *The biggest secret: The book that will change the world.* Ryde, Isle of Wight: Bridge of Love Publications.

Lalonde, J. K., J. I. Hudson, R. A. Gigante, and H. G. Pope. 2001. Canadian and American psychiatrists' attitudes toward dissociative disorders diagnoses. *Canadian Journal of Psychiatry: Revue Canadienne de Psychiatrie* 45 (5): 407–12.

Ovid. 2004. *Metamorphoses.* New York: Penguin Classics.

Pope, H. G., P. S. Oliva, J. I. Hudson, J. A. Bodkin, and A. J. Gruber. 1999. Attitudes toward DSM-IV dissociative disorders diagnoses among board-certified American psychiatrists. *The American Journal of Psychiatry* 156 (2): 321–23.

SPIRIT MEDIUMS

Spirit mediums are individuals with the ability or sensitivity to be able to contact and converse with the dead. They act as a conduit, or medium, by which the living may interact with the spirit realm. Medium activity goes back to the ancient world of oracles, shamans, medicine men, wise women, channelers, and other mystics. Of these ancients, possibly the most well known today is the Greek Oracle of Delphi. Supplicants paid a fee to the oracle (which means "one who sees"), asked a question, and the oracle would reply. The oracle, usually a woman, would go into a trance, consult with spirits and then give the waiting supplicant an answer usually couched in a riddle or other allegoric verse. The individual would then be left to interpret the response any way he wished. This basic format has been followed by mediums across time and cultures to the present day.

In the modern Western tradition, spirit mediums are most associated with the Spiritualist movement, which began in the early 19th century. In the 1840s in New York, the Fox sisters claimed to be able to contact the spirit world. Their story spread quickly. They gained supporters and adherents and the modern spiritualist movement was born. The idea spread rapidly throughout Anglo-American culture and has continued to this day. Spirit mediums fit nicely with the growing alternative religious movement then underway in America, where large portions of the population, while still devoutly religious, had grown disillusioned with the state of organized religion and were looking for something more. It also, ironically, meshed well with the growing authority of materialist science.

The advent of the Industrial Revolution brought many advances in communications technology and the understanding of the nature of electricity. It seemed logical that, if electricity could travel across a wire to transmit messages via telegraph, then why not propose a similar scientific explanation for the ability to communicate with the departed through a kind of mental telegraph. Adding weight to this concept, by the late 19th century, wireless telegraphy was being developed where it was possible to communicate over the air alone. As a result of the growing popularity of spiritualism, a series of beliefs, rituals, and standard methods soon formed in order for enthusiasts to practice it.

The central act of the spirit medium is the séance. The first book in the West on séances was Sir George Lyttelton and Elizabeth Montagu's *Dialogues with the Dead* (1760). A séance, from the French word for sitting, is a group activity where usually no more than five or six people at a time sit around a table linking hands with a spirit medium and attempt to contact the dead. Such activities go back to antiquity. During a séance, the medium goes into a self-induced trance and attempts to contact a spirit. This spirit is either specifically asked for by a participant—often a loved one—or is a spirit that appears at random. The medium is taken over by the spirit who then communicates through automatic writing and automatic speaking. In other words, any writing or speaking the medium performs is thought to originate with the spirit rather than the medium. During this process, the medium's consciousness and character steps to the back, and is often unaware of what is happening, while the spirit operates the medium's body much like a puppeteer. Spirit writing could also be performed through the use of a device with a pencil attached to it called a *planchette*. A similar device had been in use for divination for centuries. Members of the séance would lightly touch their fingers to it and the spirit would then push the *planchette* around writing on a piece of paper or pointing to preprinted words and letters already on the sheet. Commercially produced *planchettes* had been available for purchase for many years, but when a pair of American entrepreneurs intent on making money off of the spiritualist craze developed a mass-marketed *planchette* with a preprinted writing surface, the modern *Ouija* board was born.

One of the first celebrity mediums was Cora Hatch (1840–1923). When she was a child, her family attempted to join a Universalist commune in Massachusetts, but when it proved too crowded they moved to Wisconsin to begin their own. It was there in the early 1850s that the teenaged Cora discovered her gifts and her parents sent her on tour. Over the course of her life she married at least four times, usually to older men who exploited her for her revenue-generating abilities. Working as a spirit medium was a lucrative business for the top echelon of practitioners. They developed dramatic performances and touring stage acts very similar to the magic acts then popular. Ghosts appeared, objects flew about the room, and disembodied voices and music were often heard. One of the most anticipated of the effects was when the medium produced ectoplasm, a white frothy substance said to be a manifestation of psychic energy.

Some of the most successful mediums were acts like the Davenport Brothers, who performed before large audiences in theater halls. Ira Erastus Davenport (1839–1911) and William Henry Davenport (1841–1877), inspired by the Fox sisters, began experiencing the same effects of being able to contact spirits that the sisters had. Just before the American Civil War, they and their manager father took to the road to perform. They put together a device called the spirit box: a wooden contraption large enough for both brothers to enter. Volunteers from the audience would tie them up in full view and lock the brothers inside. After a moment or two, the classic séance effects would begin. Skeptics, including Jean Eugène Robert-Houdin (1805–1871), considered the father of modern stage magic and inspiration for the American Harry Houdini, investigated the Davenport brothers and found them to be frauds. Like so many spirit mediums, the Davenports were skilled performers who could make their act look paranormal. Following his inspiration's lead Harry Houdini himself famously exposed mediums and séances in the early 20th century.

A number of academics and scientists began to investigate and study the paranormal as a result of the Spiritualist movement. The English chemist and physicist William Crookes (1832–1919) had sympathy for spiritualism, but also felt it the duty of a scientist to examine paranormal claims. He released a report of his investigations in 1874 in which he said there were aspects of spirit medium practices that could not be explained by fraud or normal science. This did not stop debunkers and skeptics. One Scottish woman, Helen Duncan (1898–1956), was arrested for fraud when an undercover policewoman caught her faking a spirit manifestation in 1933. In addition to other offenses, Duncan was charged under the Witchcraft Act of 1735, much to the dismay of supporters of the paranormal and the general public alike. The uproar over the case—and the use of a law that was by then almost 200 years old—led to the abolishment of the Witchcraft Act in 1951.

There is one particularly interesting class aspect surrounding spirit mediums. The 19th-century Spiritualist movement and occult revival were largely a middle-class phenomena. They allowed respectable white women to break free of the strict cultural restraints of Victorian and Gilded Age social mores without being too ostentatious. When poor or non-Caucasian women performed essentially the same services, they were not known by the label "spirit medium," but instead by the derogatory and less respectable labels of gypsy, fortune-teller, and wise woman, occult priestess, or witch. Such women would not openly be allowed into the drawing rooms of the well off the way a white woman would.

One of the most influential spirit mediums of the post-Spiritualist craze was American Edgar Cayce (1877–1945). Known as the "sleeping prophet," he would go into a deep trance, converse with the dead, and then write down predictions for the future based upon those discussions. An early mass media medium was Brazilian Francisco Cândido Xavier (1910–2002). He appeared regularly on Brazilian television and claimed to be the reincarnation of an ancient Roman Senator, among others. A 21st-century version of Cayce and Cândido Xavier is James Van Praagh, who has made many television appearances. Another celebrity medium is John Edward McGee Jr. Known simply as John Edward, he rose to prominence because of his television show *Crossing Over* (1999–2004), on which he contacted deceased loved ones of audience members. Edward played to a large studio audience asking questions and reading them, that is, getting a sense for the spirits present and what questions audience members wished to ask. He was roundly criticized for using the non-supernatural method of cold reading. This is where a series of random, but general questions are asked and common names shouted out. With a large audience, most of who were there because they believed they could contact dead loved ones, someone was likely to recognize a name or respond to a scenario. Vague statements like "she says 'I'm alright, don't worry'" or "It wasn't your fault!" could be interpreted by almost anyone as having personal meaning. Once someone responded, a cold reader can then carefully play off the initial question or response with follow-up statements leading the respondent to some answer the cold reader could take credit for. Debunkers examining the raw *Crossing Over* tapes, which sometimes went to eight hours per show, argued that they exhibited a general lack of successful hits with the performance. Editing, however, made it seem like John Edward had produced a steady stream of successes. He was satirically skewered for using cold reading by the television cartoon *South Park*.

Not all spirit mediums are frauds and scam artists; many are sincere and genuinely believe they have the ability to contact the dead. Unlike the famous and celebrity mediums, these people rarely earn large sums for their work; in fact most barely earn enough money to survive. They do what they do because they think they are helping people. Unfortunately, sincerity is not proof of physical phenomena. Many of the most well-known genuine mediums were women who lived tragic lives, exploited by their families and husbands. Many descended into alcohol abuse. If contact with the dead was possible, it would seem that it would occur on a regular basis to almost everyone and not in ambiguous terms to a select few. It would not be considered paranormal, but simply normal, a confirmed part of daily human experience.

See also: Debunkers; Ectoplasm; Metaphysics.

Further Reading

Braude, Ann. 2001. *Radical spirits: Spiritualism and women's rights in nineteenth-century America*. Bloomington: Indiana University Press.

Nichols, Thomas Low. 2007. *Biography of the brothers Davenport*. Whitefish, MT: Kessinger Publishing.

Price, Harry. 2003. *Revelations of a spirit medium*. Whitefish, MT: Kessinger Publishing.

SPIRIT ORBS

Spirit orbs are spherical images that are invisible to the naked eye but which appear in photographs of haunted locations. They are believed variously to be energy emitted by spirits, vehicles by which disembodied spirits travel, or the souls of the departed. They are seen to float and often contain faces. The term was first applied to paranormal investigations around 1994. While most respectable paranormal investigators accept that the vast majority of orbs are dust and water particles moving on account of air currents, they also believe that a small fraction are genuine spirit entities. Paranormal author Jane McCarthy says, "This is pure speculation," but still believes that some orbs are controlled by an intelligence and have a fundamental connection to both the spirit world and UFOs. The International Ghost Hunter Society takes a slightly harder line by claiming that they have proven that spirit orbs are "the soul of a departed person." Just how they arrived at such a momentous discovery, and just what the physics of a spirit manifestation are, they do not say. They do take an unusually pragmatic step of instructing their members and would-be ghost hunters to take as many photos of orbs as possible so that a database can be established to discern genuine orbs from common dust particles.

Spirit orbs seem to exhibit no behaviors different from dust particles. The faces sometimes seen in them are simulacra of basic human visages. If orbs are something other than dust particles, there is as of yet no way to determine what they are. As with so much in the field of paranormal investigation, clear and precise definitions are crucial yet hard to come by. For example, how can it be claimed that an orb is the soul of a departed human when no acceptable scientific explanation of what a soul is exists? The English word "soul" seems to be a combination of early Germanic concepts of a life force with Greek words for vital breath. A similar concept appears in most of the world's religions in some form.

Spectral entity or speck of dust. Spirit orbs are believed by some to contain smiling faces.

See also: Ghost Hunting; Metaphysics.

Further Reading

McCarthy, Jane S. 2008. *The connection between the energy lines, the orb phenomenon, dimensions & UFOs.* n.p.: Psychic Investigators.

SPOON BENDING

The paranormal ability to bend small metal objects, often keys and spoons, with the power of the mind. Spoon bending was made popular in the 1970s by Israeli paranormalist Uri Geller who went on to have a colorful career bending spoons on television. Despite his protestations that he had genuine paranormal abilities, his nemesis, James Randi, along with others, has shown that Geller's powers are just a stage act.

Born in Tel Aviv, Uri Geller served in the Israeli Army, seeing action and being wounded in the Six Day War of 1967. After leaving the service. he began performing in Tel Aviv clubs developing what was essentially a magic act. During the 1970s. he gained fame for what he claimed was the psychic ability to bend objects like keys and spoons with the power of his mind. His powers earned him many television appearances around the world and he became a sensation. Geller claimed his abilities were paranormal and given to him by extraterrestrials, but he was dogged by accusations of fraud from the start. Magicians like James Randi and others have shown how the spoon-bending trick is done. The late Nobel physicist Richard Feynman, an amateur magician, reported that, in an encounter with Geller, he asked him to bend a spoon, but Geller was unable to do so with Feynman watching him closely.

Geller's most notorious television appearance was in 1973 when he went on the *Tonight Show* with host Johnny Carson. An avid amateur magician, Carson (1925–2002) was skeptical of Geller's claims and wanted to test him. He contacted James Randi for advice and Randi told him to set up the props in such a way that neither Geller nor his assistants could get near them ahead of time. Randi, who always believed Geller a fraud, insisted that Geller had the spoons prebent and used various tricks to read people's minds. When Geller took the stage of the *Tonight Show,* he was confronted with a tabletop of spoons, keys, and sealed metal canisters. He immediately became agitated and intimated he was being set up. Carson sat stone faced and simply asked Geller to perform the feats he was famous for: He would not. What is most amazing about Geller's performance, or lack of, on the *Tonight Show* was that it did not destroy his career. Despite being totally thwarted by Carson and Randi's simple control process, he continued on as if nothing had happened. He continues to perform and people believe in what he is doing.

Uri Geller has sued many people and organizations over the years who questioned his spoon-bending abilities. The latest, in 2007 and 2008 suits, sought to force the Web site YouTube to remove the numerous clips of him cheating that could be found there. The suits have met with mixed results.

See also: Debunkers; Ghost Hunting.

Further Reading

Harris, Ben. 1985. *Gellerism revealed.* Micky Hades International.

Randi, James. 1982. *The truth about Uri Geller.* New York: Prometheus Books.

STIGMATA

Stigmata refers to the presence of inexplicable pain and/or wounds of the body in places similar to wounds associated with Jesus Christ's crucifixion. Stigmata most notably include nail marks in the wrist or hands and feet or ankles. Other marks of this condition include wounds on the head corresponding with the crown of thorns, a single mark on the forehead, abrasions on the shoulder or chest as if from carrying the cross, whip marks on the back, bleeding from the fingers, and a puncture in the side, as if from a lance in accordance with John 19:34. Blood may also be exuded in the form of tears or sweat. The nature of these wounds can be anything from open bleeding to rash-like blotches. Occasionally, such lesions take the shape of small crosses themselves and may exude light or a floral perfume odor. The duration of this condition ranges from years of constant open wounds to recurring brief sessions of apparition; many stigmatics are noted for the clean and uninfected nature of these wounds. This phenomenon is generally most associated with Roman Catholicism

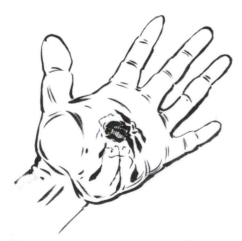

The stigmata are wounds said to replicate those of Christ at the crucifixion.

but may occur with believers from Protestant denominations; it is less common among the Eastern Orthodox. Followers of other religions have, in rare instances, exhibited similar spiritual wounds. The wounds are sometimes accompanied with claims of ecstatic visions or voices and other paranormal phenomena, such as bilocation, levitation, and extraordinarily warm body temperatures.

Medical explanations for stigmata range from psychosomatic causes to more deliberate acts of self-harm. Historic debunkers of stigmata have cited sexist evidence of it being hysteria, given its more frequent occurrence among females. Nonreligious instances of similar wounds have occurred where people had lesions from presently healed wounds. In 1928, the German doctor Alfred Lecher claimed to have unethically and deliberately used hypnosis to induce stigmata on a disturbed patient in his care. The oft-concurring religious ecstasy has been likewise attributed to self-hypnosis. Wounds frequently correspond with a religious image of the stigmatic's devotion. Ian Wilson has theorized that multiple personality disorder might enable someone to self-inflict stigmatic symptoms while in a disassociated state. Jim Schnabel linked stigmata to Munchausen's syndrome and other conditions similar to those who claim extraterrestrial abduction. The Catholic Church acknowledges the possibility of counterfeit human and perhaps demonic stigmata. Authentication by the Vatican is rare, unrelated to canonization, and requires medical as well as character analysis.

There is debatable evidence from the New Testament that the evangelist St. Paul may have exhibited this tendency, as noted in Galatians 6:17. The first prominent example was the case of the 13th-century friar St. Francis of Assisi. The most prominent recent example of this occurred with St. Pio of Pietreclina, also known as Padre Pio (1887–1968), a member of a Franciscan order. Professor A. Bignami, a pathologist investigating the Italian friar in 1918, found traces of iodine in his stigmatic wounds, plausibly indicating self-infliction. Controversy regarding Pio's condition reached the Vatican and, upon his death, the wounds had allegedly vanished without scars. The most noted living stigmatic today is likely the Catholic priest Zlatko Sudac of Krk, Croatia.

Further Reading

Harrison, Ted. 1994. *Stigmata: A medieval mystery in a modern age.* New York: St. Martin's.

Nickell, Joe. 1993. *Looking for a miracle: Weeping icons, relics, stigmata, visions & healing cures.* New York: Prometheus.

STONEHENGE

An ancient, megalithic structure found in Wilshire, England, near Salisbury thought by archaeologists to have been built during the Neolithic Age, Stonehenge went through a slow metamorphosis over time. Starting out as an earthen mound with an encircling ditch, it was a burial site. It has gone through a series of phases with a wooden post ring, associated roads and avenues, and then various arrangements of standing stones until it reached the form known today. Its earliest appearance in the written record can be found in the writings of Julius Caesar who attributed it, incorrectly, to a religious sect known as Druids (neodruids still perform rituals at the site). It was alternately a cemetery, ritual site, astronomical observatory, and a kind of Stone Age community center. The site is currently made up of large standing stones

with crosspieces mounted on top. This arrangement is somewhat in question, as late-19th- and early-20th-century researchers moved stones around, stood up stones that were lying down, and kept imperfect records of what they did. The current version, known around the world as the symbol of ancient monuments whose popularity is second only to the pyramids of Egypt, dates to about 2500 BC, but may go back to 3000 BC.

Because of its great age and unusual construction, Stonehenge is of interest to archaeologists studying the Neolithic British past. Extensive work done by scientists and historians beginning in the later half of the 20th century has discerned much about the site and the people who built it, and excavations have turned up many artifacts and human burials. This shows that the structure, at least during some periods, was a heavily populated place. It has also been shown that Neolithic people with only a crude technology could have erected even the largest stones on the site with relative ease. Some esoteric authors argued that it would have been impossible for such people to cut, transport, and raise the stones by themselves; this led to speculation that the stones were erected using antigravity devices, lost knowledge, or with the help of extraterrestrials.

After Caesar, an early mention of Stonehenge appears in Geoffrey of Monmouth's *History of the Kings of England* (c. 1135). He wrote that the stones used to build Stonehenge were brought from Africa to Ireland by a race of giants because the stones had magical powers. The magician Merlin then transported the stones to their current site. Pagan revivalists began holding ceremonies at the site in the 1870s. The stones were once completely open to the public, but changes were made in the 1980s, particularly after

Generally accepted as a religious site, Stonehenge, with its megalithic stone architecture, is a source of much fantastic speculation.

the "Battle of the Beanfield" in 1985. A group of New Age enthusiasts wanted to hold a festival on the site but local authorities tried to stop them. Violence erupted, with the revelers accusing the police of starting the violence and the police claiming it was the visitors who started it. Since this incident, access to the site has been restricted, and while neopagans are allowed to hold ceremonies at certain times of the year, usually during the solstices, these too are carefully controlled.

Both archaeologists and armchair investigators regard Stonehenge as an astronomical site. Alignments are seen that have the sun rising and setting over specific spots in the structure and it seems clear that these phenomena were planned in advance. This suggests that the site's builders had some knowledge of astronomy. While the careful work of archaeologists, anthropologists, and historians has helped put to rest some of the wilder claims about Stonehenge, such specula-

tion continues. In 2002, claims were made that Masonic and occult wrongdoing in the form of "Black-Occult" rituals were being conducted there by members of the local police and the British military. As with all megalithic sites around the world, Stonehenge has been associated with UFOs, Atlantis, and secret societies. Authors regularly claim to have solved the meaning of Stonehenge, but archaeologists argue that there was no one meaning to the site. Meaning and use changed over time as different peoples employed the site, added to it, changed it, and reimagined its purpose.

See also: Hidden History; Ley Lines; Unidentified Flying Objects (UFOs).

Further Reading

Johnson, Anthony. 2008. *Solving Stonehenge: the key to an ancient enigma.* London: Thames & Hudson.

North, John. 2007. *Stonehenge.* New York: Free Press.

TESLA, NIKOLA

A pioneering electrical engineer and systems designer, Nikola Tesla (1856–1943) designed the first practical alternating current (AC) system, introduced florescent lighting, made early breakthroughs in radio technology, foresaw the use of robots, radio control, and a host of other innovations. He worked on a wide range of ideas concerning the application of electricity, particularly wireless technology. His prodigious intellect visualized a host of devices that are still being used and explored in the early 21st century. Little discussion of the history of electrical engineering is possible without reference to him and his work.

Ethnically Serbian, Tesla was born in Croatia and, at the prodding of his illiterate peasant mother, read technical books and memorized them. He developed a powerful ability to visualize abstract ideas in his head. This ability would often cause his future colleagues frustration, as he would have entire systems worked out in his mind but not on paper. His intellect did have its cost. He suffered from headaches, often saw lights flashing before his eyes, and may have suffered from attention deficit syndrome and obsessive-compulsive disorder. He received most of an intensive engineering education, but was forced to drop out before graduating because of financial difficulties. In the 1870s, he found work at the Paris offices of the Edison Company. He impressed his superiors who suggested he go to the New World to find his destiny; in 1884, he went with little more than a few articles he had written, a few dollars in his pocket, and many dreams.

The one major thing Tesla brought with him was an idea for an alternating current motor. AC was already known and a number of engineers were working on designs but no one had been able to make a practical model. Tesla felt he had worked it out. He went to the Edison Company offices in the United States and there, too, quickly impressed people. Edison hired him to work on electrical generators, called dynamos. He did excellent work but soon found himself at odds with Edison. The two men were very different. Edison was a self-taught inventor rather than an engineer or a scientist. He was crude, slovenly, tyrannical toward his underlings, and his lack of theoretical scientific training meant he doggedly tried one idea after another until it worked, instead of thinking past the problem. In contrast, Tesla was a sophisticated, erudite theorist (though he too had a mechanic's natural ability to fix machinery) who knew a little theory could go a long way to eliminate wasted effort. The two men clashed often, and when Edison arrogantly cheated Tesla out of royalties he had been promised, he quit and went to work for Edison's rival George Westinghouse. It was at the Westinghouse Company that Tesla finally brought AC into its own. This revolutionized electrical engineering and allowed for the electrification of the world.

Tesla's eccentric character tended to make people see him as an oddity. That he made unusual pronouncements about things he was working on or how things would be in the future made many wonder. Tales of robot armies, flying machines, and electrical power intrigued but baffled the public. He reveled in performing apparent death-defying tricks like allowing high levels of electricity to pass through his body in order to operate lightbulbs, or sitting calmly in a laboratory as enormous sparks of electricity crackled and shot around him. His entire persona

Nikola Tesla, the tragic engineering genius, pioneered radio, robotics, florescent lighting, and a host of other technologies. An early biography called him an alien from Venus.

led some esoteric writers to claim he had magical powers, was in contact with aliens, or even was an alien.

One of Tesla's boldest and most controversial assertions was that he knew how to extract energy from the earth and the surrounding atmosphere in such a way that it would provide free energy for everyone, anywhere on earth. This caused competitors like Thomas Edison and even backers like J.P. Morgan to question his sanity. The idea of free energy was appealing to the average person, but not to the energy industry. Morgan dropped his backing and others helped to ruin his image. Tesla was already thought of as peculiar, now he was seen as dangerous. Opponents claimed he was working on energy weapons that would put America in peril. He himself had made the argument that energy could be used as a "death ray." This led the media to portray him as a dangerous genius, and

he became the modern high-tech model of the mad scientist. Tesla, however, was not using his intellect to do harm, nor was he doing anything occult or supernatural. He was simply following his own imagination to expand science in new ways. He was, in fact, a humanitarian who sought to use science to end war and for the betterment of all people. In this way, he was a utopian dreamer as well as scientific pioneer.

The life and career of Nikola Tesla is an example of how advanced science can be perceived as pseudoscience and how it can harm a career. Furthermore, Tesla is an example of how, when science or technology is misunderstood, it can be seen as magical. Without any way to explain how Tesla did some of the things he did, or explain some of the visions of the future he had, skeptics at the time, and since, had no choice but to label his work supernatural and him a crackpot. Just because an effect cannot be adequately explained at the moment does not mean there is no explanation. It does not allow us to fall back on supernatural explanations. Tesla's life forces us to understand that learning about the universe is not always easy. It takes hard work, and then we still may not understand. We must be careful, then, not to succumb to the easy way out and put things down to fairies, aliens, or God.

Further Reading

Childress, David Hatcher, ed. 1993. *The fantastic inventions of Nikola Tesla.* Kempton, IL: Adventures Unlimited Press.

Jones, Jill. 2004. *Empires of light: Edison, Tesla, Westinghouse, and the race to electrify the world.* New York: Random House Trade Paperbacks.

Regal, Brian. 2005. *Radio: The life story of a technology.* Westport, CT: Greenwood.

Seifer, Marc. 2001. *Wizard: The life and times of Nikola Tesla: Biography of a genius.* Secaucus, NJ: Citadel Press.

TUNGUSKA EVENT

The Tunguska Event was a devastating and perplexing explosion that occurred in central Siberia in 1908. It laid waste to 1,000 square miles of forest, was heard in western Europe, and caused atmospheric effects as far away as London. Earthquake recorders around the world registered the vibrations. Subsequent investigations of the event have led to various explanations for it including a comet's impact, a nuclear explosion, and the crash of a UFO. Standard wisdom today holds that the likely culprit was the strike of a stony or iron meteor.

On June 30, 1908 at 7:15 A.M., a titanic explosion rocked central Siberia east of the city of St. Petersburg, Russia. It was not until 1927 that a Russian scientific expedition led by mineralogist Leonid Kulik (1883–1942) first traveled to the site to investigate. Kulik discovered a massive area of damage radiating outward from a central point near the Tunguska River. Baffled but determined, Kulik returned to the spot throughout the 1930s and became convinced that it was the site of a meteorite that had crashed into the earth. (A meteor is an iron body that enters the earth's atmosphere from space and explodes above the surface, while a meteorite is one that survives an impact with the surface of the earth.) Kulik searched what he thought was an impact crater, but which turned out to be a bog common to the region. In 1938, he had aerial photographs taken that showed trees had been knocked down in a butterfly pattern. When World War II began, Kulik fought in the defense of Moscow, was wounded, and captured by the Nazis. He died in a prisoner-of-war camp.

A series of Soviet scientists later took up the cause of discovering the source of the Tunguska explosion, as have Americans and Italians. They have put forward a range of explanations, including meteorites and comets. Less plausible are black holes and "natural H-bombs." More outlandish is the suggestion that a UFO crashed and exploded. Soviet engineer and science fiction writer Alexander Kazantsev (1906–2002) argued after seeing the devastation at Hiroshima that a UFO searching for water at Russia's Lake Baikal crashed and set off a nuclear explosion. Other fantastic explanations claim the event was the result of experiments by Nikola Tesla and that the explosion was caused by the release of pent-up natural gases and geothermal energy from within the earth. Russian Astronomer Igor Zotkin conducted a series of laboratory tests with a simulation of the Tunguska area and saw that an airburst type of explosion gave a damage pattern to model trees in the same butterfly shape that Kulik saw.

In the end, the most plausible explanation for the Tunguska event is that either an iron meteor or a stony meteor (a broken piece of an asteroid) entered the earth's atmosphere on June 30, 1908, and exploded in the air over Siberia. Scientists point out that, if such an event had happened, or does happen, over a major urban center it would have apocalyptic results and would no doubt change human history, if not mark the beginning of its end.

Further Reading

Lerman, J.C., W.G. Mook, and J.C. Vogel. 1967. Effect of the Tunguska meteor and sunspots on radiocarbon in tree rings. *Nature* 216:990–991.

Verma, Surendra. 2006. *The mystery of the Tunguska fireball*. London: Icon Books.

UNIDENTIFIED FLYING OBJECTS (UFOs)

UFOs are alleged conveyances from other worlds that regularly appear in the earth's atmosphere. The term was coined by the U.S. Air Force in 1952 to describe any aerial phenomena that has not been identified, though the term is most commonly connected to nonterrestrial or alien spacecraft. Sentient alien beings are said to use UFOs to visit the earth for various reasons.

The majority of UFOs are simply sightings of unusual lights in the sky. Occasionally they appear as solid objects known as daylight disks. They have been photographed and filmed, though such films are usually fleeting in nature, and whether moving or still, the images are often seen at a distance with little or no surface detail discernable. Ufologists (those who study UFOs) acknowledge that the majority of sightings are misidentifications of known phenomena and aircraft and that a number of sightings and photos are intentional hoaxes.

Some researchers see alien visitations as a far from modern development. They cite legends of gods, beings of light, and similar occurrences in mythology and scriptures from cultures worldwide to show these visits have occurred throughout human history. From Ezekiel's wheel in the Bible to the Bimana of Vedic, scriptures have been held up as evidence of early encounters with UFOs. Believers also point to pictorial representations from cave art to the medieval period of objects that appear to conform to modern descriptions of alien craft. During the Second World War, Allied aircrews reported strange balls of light that seemed to follow their planes, dance around them, but not interfere with them in any way. It is alternately argued that pilots took the name for these strange lights, Foo Fighters, either from the French word for fire or from the antics of a popular cartoon character.

The modern UFO era began June 24, 1947, when American businessman Kenneth Arnold saw a group of crescent-shaped craft near Mount Rainier, Washington. As he flew a small private plane, he watched a formation of silvery objects fly by him at great speed and in a way he said resembled a saucer skipping across water. Hearing Arnold's description, a local newspaper man coined the term *flying saucer*. The term flying disk was also used. Suddenly, people around the world were seeing all sorts of strange things in the sky and speculating as to what they might be.

There are a number of explanations used to account for the nature of UFOs. The most common is the extraterrestrial hypothesis (ETH), which states that UFOs are nuts and bolts technology used by alien civilizations to travel to and inspect the earth. This idea was first stated more roughly by anomalist writer Charles Fort in *The Book of the Damned* (1919), but coined by University of Colorado physicist Edward Condon as part of his *Condon Report* on UFOs in the 1960s. The interdimensional hypothesis also seeks to explain UFOs, suggesting that UFOs are from a different physical dimension here on earth. A number of paranormal,

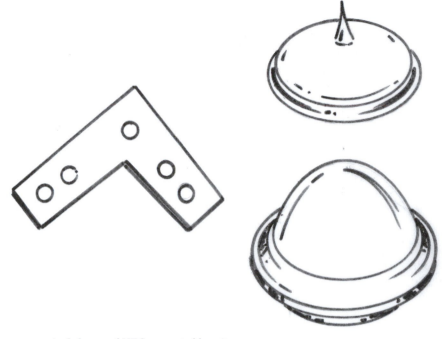

These are typical shapes of UFOs reported by witnesses.

spiritual, or occult explanations have also been put forth. While there are individual scientists who believe UFOs are genuine, there is no physical evidence of them accepted by mainstream science. The bulk of the evidence comes in the form of eyewitness accounts, which run the gamut from those made by trustworthy observers such as military and police personnel to those which are obviously hoaxes or are made by individuals with ulterior motives. There are also photos and film footage, and, the least numerous form of evidence, traces left behind on the ground.

An early supporter of the ETH was aviation writer and former Marine Corps Major Donald Keyhoe (1897–1988). He was the author of some of the first and most influential UFO writings beginning with the article "Flying Saucers are Real"

in the popular men's magazine *True*, published in January 1950. This was quickly followed by a book-length version of the article, called *The Flying Saucers Are Real*, which was released in the same year, and another, *Flying Saucers from Outer Space* (1953). Keyhoe argued that UFOs had been coming to earth to observe human behavior for several centuries. He made the connection between atomic bomb testing and the sudden increase in UFO activity. He also suggested that these beings posed no threat to humans. His scenario worked its way into many popular film depictions of aliens both indirectly, as in *The Day the Earth Stood Still* (1951), and directly when he acted as technical advisor for *Earth Versus the Flying Saucers* (1956).

In August 1947, just one month after the Arnold sighting, the now-infamous

Roswell incident occurred. The U.S. military at first announced and then denied a UFO had crashed in the New Mexico desert. By the following January, the Air Force had quietly begun Project Sign to investigate the UFO phenomenon. Members of the project came to support the ETH view as a probable explanation. Air Force General Hoyt Vandenberg then removed those members and reconstituted the operation into Project Grudge. This new enterprise had a distinct anti-UFO bias. The official Air Force position was that there was nothing to the ETH. By 1952, a new and more public Air Force operation called Blue Book was announced. Critics charged that Blue Book was really nothing more than an attempt by the Air Force to prove UFOs were not real and to ridicule those who believed they were. Blue Book officially closed down in 1969, after Air Force officials stated there was nothing left for them to do, as they had shown claims of UFOs and the ETH to be groundless.

One of the first scientific investigators of UFOs was the astronomer J. Allen Hynek (1910–1986). He was asked by the U.S. Air Force to serve as a civilian consultant for their three most widely known UFO investigations projects from 1947–1969. He came to believe that some UFO reports were indeed of extraterrestrial craft. In order to categorize these sightings, he developed a hierarchy of "close encounters" that has since become widely applied. A close encounter of the first kind (CE1) described sightings of UFOs in the viewer's relative vicinity that had no interaction with people or the environment. A close encounter of the second kind (CE2) described UFOs that left some evidence of their presence upon the environment, like marks on the ground or traces of radioactivity. The

final form, a close encounter of the third kind (CE3) was often a combination of the first two along with direct observation of and interaction with the occupants of the craft.

Unidentified flying objects have been sighted around the world. While it has been common to assign specifically American cultural significance to UFOs, because they often represented postwar worries about nuclear war, the spread of communism, and other Cold War concerns, the international cross-cultural and cross-chronological nature of the phenomena belies such easy meta-explanations, however UFOs do often take on specific cultural significance relative to the places in which they are seen. For example, a major "flap" (group of sightings) that occurred in Mexico in the 1990s had special local religious and prophetic significance. Descriptions of the occupants of UFOs also vary depending on the culture observing them. Western European and Anglo-American sightings tend toward the Grey aliens, Russian sightings often include robotic characters, and Latin American sightings often include reptilian creatures. Insect-like creatures and humanoids are also sometimes seen.

UFOs are studied anecdotally by viewing video footage, still images, and, most often, reviewing eyewitness accounts. Like ghost hunting, the search for UFOs is made problematic by a lack of indisputable evidence and research methodology. Without knowing what UFOs really are, it is difficult to establish effective research techniques. Of the various residual phenomena reported to be associated with UFO encounters—burned ground areas, raised levels of radioactivity, cattle ripping, and unusual lighting effects—all have alternate, prosaic explanations which could account for them. As of yet, there are no examples of residual debris or ground effects that have been unquestionably attributed to extraterrestrial causes. There is also a lack of uniformity to UFO evidence. Though there are examples of similar objects seen in different parts of the world, photographs and films regularly show very different types of craft. There is also a lack of uniform residual evidence left behind by reputed UFO landings.

As with so much in the world of pseudoscience, UFOs are associated with a great deal of circumstantial and intriguing evidence, but nothing concrete. Given that science works by analyzing physical evidence, there has been disappointingly little to study despite worldwide belief and interest in the subject, even on the part of academic scientists. While the mainstream scientific community does believe the chances are good that sentient life exists elsewhere in our universe, few accept the idea that any of those possible life forms have visited the earth now or in the past.

See also: Alien Abduction; Alien Autopsy; Fort, Charles Hoy; Roswell Incident.

Further Reading

Evans, Hillary, and Dennis Stacy, eds. 1997. *UFO 1947–1997: Fifty years of flying saucers*. London: John Brown.

Yenne, Bill. 1997. *UFO evaluating the evidence*. New York: Smith Mark.

VAMPIRES

Vampires are re-animated dead humans who live on the blood of the living, primarily humans but occasionally animals. Vampires drink the blood of their victims by biting them to initiate the blood flow, and then imbibe it. There are vampire-like legends from most cultures around the world. The modern word "vampire" comes from folklore prevalent in the Balkans and southeastern Europe, and became popular in the 18th century. The folkloric vampire is a rotted, bloated, slow-moving, and repulsive thing that preys upon victims local to where it was originally buried. Details of vampiric behavior vary according to their home cultures.

Vampires get their cultural punch not from historical sources, but from literature. Most of the accoutrements of vampire lore are inventions of novelists and filmmakers. The first modern vampire novel was John Polidori's *The Vampyre* (1819). Polidori wrote his story while staying in the company of the poet Percy Bysshe Shelley and his wife Mary Wollstonecraft Shelley as they toured Europe in the summer of 1816. It was during this same period that Mary Shelley began writing *Frankenstein* (1818). The ultimate vampire novel was published later that century when Irish playwright Bram Stoker (1847–1912) published *Dracula* (1897). Stoker based the character of Count Dracula on the historical figure of Vlad Tepes (1413–1476), also known as Vlad the Impaler and, in his native Wallachia/Transylvania area, called Vlad Dracul (son of the dragon). While Vlad was a ruthless leader and mass murderer, he did not drink his victims' blood. Stoker combined this character with legendary ideas about vampires to create Dracula and then turned a loathsome fiend into a suave and debonair socialite. The modern romantic image of the sexy vampire is as far from the folkloric version as can be imagined. What most people believe are ancient vampiric tendencies, as with those of the werewolf, are mostly inventions of Hollywood. Vampires are still believed to exist in places around the world. Vampire panics were reported in Africa as late as 2005.

See also: Werewolves.

Further Reading

McNally, Raymond T. 1994. *In search of Dracula: The history of Dracula and vampires.* New York: Mariner Books.

VIMANA

Vimana are flying machines described in a number of ancient Indian Sanskrit poems and epic literature including the *Vedas, Ramayana,* and *Mahabharata.* They are flying wheeled chariots that act as conveyances for the gods. These texts date back to about 1500 BC, but likely go back further as oral traditions. Some ufologists see *vimana* as ancient descriptions of flying saucers and other extraterrestrial machinery. They are seen as the ufological equivalent to biblical descriptions of Ezekiel's wheel and others. The craze for linking the *vimana* to UFOs began with the publication in the 1970s of a supposed Vedic text called the *Vaimānika Shāstra.* The text was supposedly channeled, or recited by a dead person through a living one while in a trance. A Hindu mystic was able to contact an ancient Hindu mystic through occult means and he gave him the text orally. The text describes the flying machines, how they are made, how they are flown, and what they are used for. While impressive sounding to laymen, when read by trained engineers, the *Vaimānika Shāstra* comes off as jibberish. The *vimana* are another in a line of attempts to find historical proof of UFOs. Ufologists have been raiding the ancient texts and histories of all the world's literature looking for evidence that can be shoehorned into UFO belief, but historians know it is problematic to try to interpret past peoples and their writings in terms of modern ideas.

See also: Unidentified Flying Objects (UFOs).

VON DÄNIKEN, ERIC

Eric Von Däniken (b. 1935) is the speculative Swiss writer and hotelier best known for *Chariots of the Gods* (1968), in which he argued that archaeological evidence existed to prove intelligent beings from other worlds had visited the earth in the ancient past and heavily influenced the course of human history. He gained worldwide notoriety for his book, which helped create a major interest in the subject. His assertions have mostly been ignored by archaeologists and Egyptologists or debunked as nonsense. Von Däniken had no training in archaeology or world history and misrepresented some of his evidence. His use of a picture of an ovoid shape drawn into the ground at Nazca, Peru, which Von Däniken said looked like the parking area for large airplanes at an airport, drew criticism because it was actually only a few feet across (the photo in his book made it seem huge). Von Däniken argued that they were used by extraterrestrial spacecraft as a parking area while awaiting takeoff. His reputation was not helped by a fraud conviction and tax troubles, for which he spent time in a Swiss jail. Von Däniken later recanted some of his evidence, admitting that it was manmade and not of extraterrestrial origin as he had originally claimed. A theme park based on his work closed due to financial failure in 2006.

See also: Ancient Astronaut Theory.

VOYNICH MANUSCRIPT

The Voynich manuscript is an extraordinary medieval manuscript that has defied deciphering for centuries, even by modern military cryptologists. Written sometime in the late 15th or early 16th centuries in Europe, its inability to be read has only increased interest in the work and inspired a number of authors to attempt to explain its origins, its history, and authorship. It is most commonly attributed to Francis Bacon, though the case for Bacon is considered problematic.

The manuscript was named for Wilfrid Voynich, the rare-book dealer who first brought it to modern notice in 1912. It is thought to be the only medieval manuscript that remains without translation. The arrangement, repetition, and use of the words in the work do not seem to follow accepted ways of language writing. Also, there are no other known examples of the script used. This has led some scholars to argue that the Voynich manuscript is an elaborate hoax written in a made-up language that cannot be deciphered because there is no language to decipher, only gibberish. There are illustrations (illuminations) throughout the text that suggest it is an herbal or astronomical work of some sort. It currently resides in the Beinecke Rare Book and Manuscript Library at Yale University in New Haven, Connecticut.

WATER MEMORY

Water memory refers to the idea that ordinary water is able to retain the memory of materials dissolved into regardless of how diluted the water is rendered. In other words, if a material is diluted in water to the point that not even a single molecule of that material remains, the water will still behave as if it still had a high concentration of that material in it. This was once thought to be a significant scientific support for the alternative medical practice of homeopathy, as a great deal of homeopathy is based on the dilution of remedies in water. If proven, water memory would allow a patient to take various medical remedies in a highly diluted form without the risk of certain side effects, yet he would still gain the full recuperative powers of the remedy. Despite repeated attempts to prove the efficacy of water memory, there has been no successful, scientifically controlled, double-blind test of it accepted by the mainstream scientific and medical communities.

The main champion of water memory was French biologist and immunologist Jacques Benviniste (1935–2004), whose career was planted firmly in the mainstream. After graduating from medical school, he chose to go into immunological research instead of private practice because of a back injury he had sustained during his early career as a racecar driver. In the 1960s and 1970s, he worked first for the French Cancer Research Institute (CNRS) and then the prestigious Scripps Clinic and Research Foundation in California where he did important work on human blood structure. By the mid-1970s he had been appointed head of the National Institute of Health and Medical Research (INSERM), a French government-sponsored body. It was at INSERM that his water-memory work began. His team diluted human antibodies, yet found that the water in which they had been diluted still reacted when mixed with other materials in the way they would have had they not been diluted. It worked only when the solution was shaken vigorously. This was a startling and important discovery and promised exceptional relief for allergy sufferers.

In 1998, Benveniste collated his team's work and sent it as a paper to the highly regarded journal *Nature*. Such a publication would give the water memory idea the stamp of approval of the scientific world and, by association, support the maligned field of homeopathy. The *Nature* editorial board found itself in a quandary. They could find no obvious mistakes or faulty methods in the paper, but to accept it for publication would mean suspending belief in large chunks of the known laws of chemistry. *Nature*'s senior editor John Maddox made a compromise. Benveniste's paper was published in June 1988, but an editorial by Maddox ran with it. He stated his concerns and articulated the problems with the laws of chemistry with which water memory conflicted. To solve the conundrum, he suggested that a double-blind test be made to see if Benveniste's results could be verified. A referee team was assembled that included Maddox and magician and pseudoscience foe James Randi. Both sides agreed upon a set of rules and understandings, and the test began. The test produced results just as in the paper. Maddox, however, was suspicious of some of the ways the lab technicians could see the difference between test samples. One more test was agreed upon, and this time with the lab team sequestered while the test samples were prepared so that no one would know which test sample was which. This time, the test failed to produce the desired results.

Following the test, *Nature* ran an article with its evaluation of water memory. The editorial board said that belief in water memory was "unnecessary as it was fanciful." While *Nature* did not accuse Benveniste of fraud, they were concerned that several members of his team were on the payroll of a major French homeopathic company called Boiron. The *Nature* team argued that all the test procedures were agreed upon ahead of time and that they did nothing Benveniste and his team did not know was going to be done. The mainstream scientific community saw this as a repudiation of water memory in particular and homeopathy in general, while the homeopathy community saw the test as just one more insult and dismissal of their work and the confirmation that the mainstream was trying to crush them. Benveniste continued his experiments—despite being fired from INSERM—with the support of the homeopathic industry. Since then, a small group of scientists at other laboratories sympathetic to water memory have claimed to have replicated the positive tests results, or at least have argued there is enough to water memory to continue research. In 1997, Benveniste founded the DigiBio company funded by the homeopathy industry and dedicated to developing various applications for digital biology (what water memory was now being called).

See also: Blind Testing; Homeopathy.

Further Reading

DigiBio.com Web site.
Maddox, J., J. Randi, and W. W. Stewart. 1998. "High-dilution" experiments a delusion. *Nature* 334:287–90.

WELLCOME, HENRY

Henry Wellcome was an eccentric American drug-company entrepreneur and self-made millionaire who used his wealth to amass one of the largest collections of anthropological and medical history books, ephemera, and artifacts in the world. He began his career as an assistant in his uncle's Midwestern pharmacy and quickly worked his way up. He qualified as a pharmacist and dreamed of doing something altruistic for humankind. In 1880, he traveled to England and, with a partner, formed Burroughs Wellcome & Co. He coined the word "tabloid." Wellcome (1853–1936) traveled widely but also employed paid collectors extensively. He was especially interested in the more unusual aspects of anthropology and medical history, especially from third world countries. He collected ancient manuscripts from around the world, books, medical instruments, medical advertising, sexual aides, religious fetishes, medical furniture, and medicines themselves. He was able to reconstruct entire apothecary shops and an alchemist's laboratory. The collection eventually reached over one million pieces. In 1936, the year of his death, he established the Wellcome Trust to promote medical research and philanthropy. The bulk of the collection is located in London.

See also: Alternative Medicine.

Further Reading

Arnold, Ken, and Danielle Olsen, eds. 2003. *Medicine man: The forgotten museum of Henry Wellcome*. London: British Museum Press.

Gould, Tony, ed. 2007. *Cures and curiosities: Inside the Wellcome library*. London: Profile Books.

WEREWOLVES

Werewolves are legendary creatures that are a composite of humans and wolves. The werewolf is a human who is transformed against his will into such a creature. Generally associated with European folklore, werewolves (from a combination of Old English and Germanic words for wolf-man), or lycanthropes (from the Greek for wolf-man), are mythical figures that transform, or shape shift, into wolves or wolf-like creatures, though not always at the full moon. Similar legends are known back to ancient times. The Roman author Pliny the Elder wrote of a monstrous race of dog-headed men called the Cynocephali. They were not reputed to be shape-shifters, however. As one usually is made a werewolf by being bitten or mauled by an existing werewolf, lycanthropy is normally viewed as a curse. In some cultures, shamans (priest-like religious figures) shape shift as part of a religious ritual. Such rituals, commonly found among Native American religious beliefs, call for the shaman to don animal skins to take on the appearance of a wolf. Shamans may also imbibe hallucinogenic drugs so that they feel like wolves. The word Animagus, one who can turn themselves into a werewolf intentionally, is a fictional creation of author J. K. Rowling and is not historic.

There is no set canon of werewolf belief. Superstitions vary from one culture to the next about the meaning of lycanthropy, how one becomes a werewolf, how werewolves live, and how they die. The idea of dispatching a werewolf with silver bullets is a 20th-century cinematic invention, as is possibly the most famous piece of werewolf poetry written by screenwriter Curt Siodmak for the film *The Wolf Man* (1941):

> Even a man who is pure of heart
> And says his prayers at night
> Can become a wolf

When the wolf bane blooms
And the autumn moon
Shines bright

There are few recorded cases of trials for lycanthropy during the witch crazes of Europe in the 17th and 18th centuries. The best documented is that of an old man named Thies of Kaltenbrun, of Latvia, in 1691. The court did not take lycanthropy that seriously and certainly not this old man who said he was a werewolf and had traveled to hell and back. He made numerous fantastic claims before the bench. The court had only called him as a witness in a separate witchcraft case, and was apparently sorry they did.

The origin of the werewolf myth is uncertain and perhaps has no singular genesis, as many cultures have lupine lore. The Greek god Zeus turned Lyceaon, founder of Acadiea, into a wolf. A cult arose around that myth that lasted into the early years AD. It is from this myth that the condition of being a werewolf

got the name *lycanthropy*. Totems and battle practices of Germanic, Scandinavian, and Pictish warriors often involved wolves and have been cited as an influence on the myth, as have primordial and anthropological justifications for lupine phoebes. Lycanthropy as a term has authentic European medical connotations for one with appetite and mannerisms of a wolf. In many circumstances, the viable treatment was legal execution. Lycanthropy was correlated with witchcraft particularly in France, where it was used as an accusation in the Valais witch trials.

Sightings of werewolves are not common at present, and no large community of lycanthropic believers exists. There are communities, particularly in Africa and India, that still believe in were-creatures though they are often other than wolves. One might conjecture a correlation among these sightings and other myths of hair-covered human attackers. Lycanthropic illness may have been projected on the mentally ill in medieval and early modern Europe. Nazi Germany incorporated lycanthropic and lupine imagery into the cannon of military terminology. Evidence of ailments related to lycanthropy is rather shallow, but many theories exist as to what may have made someone be mistaken for a werewolf. Other creatures of similar morphing ailments turn to bears, boars, tigers, hares, or crocodiles.

Werewolves, like vampires, are not cryptids, but rather folkloric creations of the imagination. Humans do not have the ability to shape shift. While they may be romantic figures, werewolves are not flesh-and-blood animals.

See also: Shape Shifting.

Further Reading

Baring-Gould, Sabine. 1973. *The book of werewolves*. New York: Causeway Books.

Half man, half wolf, the werewolf is popularly depicted as a vicious but tragic figure.

De Blécourt, Willem. 2007. A journey to Hell: Reconsidering the Livonian "werewolf." *Magic, Ritual and Witchcraft* 2 (1): 49–67.

Edwards, Kathryn A., ed. 2002. *Werewolves, witches, and wandering spirits: Traditional belief and folklore in early modern Europe*. Kirksville, MO: Truman State University Press.

Godfrey, Linda S. 2008. *Werewolves (mysteries, legends, and unexplained phenomena)*. New York: Checkmark Books.

WORLDS IN COLLISION

Worlds in Collision is a controversial book by Russian author Immanuel Velikovsky (1895–1979) in which he argued a catastrophic view of the origins of the solar system, human history, and the creation of the planet Venus. Velikovsky argued his point by marshaling folkloric, legendary, and ancient cultural traditions and knowledge to argue for certain historical movements of heavenly bodies. According to *Worlds in Collision* (1950), about the 15th century BC, a large body was spontaneously ejected from the planet Jupiter. This body passed so close to the earth that it caused catastrophic meteorological and geological upheaval. It came close to the earth again about half a century later, bringing another round of apocalyptic earth destruction. It then assumed a predictable orbit around the sun and assumed the role of the planet Venus. Though widely popular, the scientific community roundly criticized *Worlds in Collision*, especially astronomers who argued that Velikovsky had badly misinterpreted the workings of celestial mechanics and that the basic ideas of the work bordered on fantasy. The book has remained controversial and has garnered a cult following.

As noted, the scientific community reacted swiftly and negatively to *Worlds in Collision* and quickly pointed out that the way Velikovsky describes the movements of the planets violates everything known about Newtonian physics. Velikovsky admitted that he was proposing that gravity did not always work the way astronomers thought. Indeed, prior to *Worlds in Collision* he published a pamphlet title *Cosmos without Gravitation* (1946). He was criticized for not understanding celestial mechanics—how planets behave—and that a number of his basic premises were posited despite evidence being discovered and published that countered him prior to the writing his book.

Well-known astronomer Carl Sagan was especially active in denouncing *Worlds in Collision* and he included an entire chapter on it in his book *Broca's Brain* (1979). He argued that everything astronomers knew about the history, atmosphere, geology, and orbiting habits of Venus contradicted Velikovsky's work. Archaeologists also denounced Velikovsky's skewing and rewriting of ancient history as amateurish and wrongheaded. Velikovsky had rearranged and reinterpreted known ancient history and genuine source materials in a quirky way to support the idea that ancient people knew of these catastrophic earth changes. He said that the two close passings of the future planet Venus and the havoc they wreaked had been recorded in different religions as stories and descriptions of their deities.

Immanuel Velikovsky's *Worlds in Collision* helped pave the way for later neocatastrophist, antiscience and archaeology writers like Zacharia Sitchin and his *The 12th Planet* (1976) and Graham Hancock's *Fingerprints of the*

Gods (1995). Sitchin hypothesized about a planet unknown to modern astronomers but well known the ancients called Marduk. He also engaged in speculation about ancient astronauts being the sources for early religious traditions. Hancock proposed that, around 5500 BC, a kind of benevolent golden-age race of people were scattered by catastrophic planetary change. The survivors traveled around the world and helped create the known ancient civilizations of Egypt, Mesoamerica, and others. Both these works, like *Worlds in Collision,* are based on quirky interpretations of archaeological knowledge by nonarchaeologists, ancient languages by nonlinguists, and astronomical movements by nonastronomers to completely rewrite human history in terms of titanic catastrophic upheavals, lost super races, and the general notion that there is a hidden human history that the scientific establishment actively works to suppress knowledge of. Even more insidiously, these works promote the idea that academic specialists who spend their careers studying a topic like archaeology, astronomy, or geology are bumbling simpletons who are blind to the truth of what the materials mean, and that these specialists cannot be trusted. It is a fear of intellectuals that promotes the idea that an amateur investigator working with a complex subject like planetary mechanics, Babylonian history, or the ancient Sumerian language is more trustworthy than the professional scholar. This makes *Worlds in Collision* and its progeny the poster children for pseudoscience and pseudohistory.

See also: Hidden History; Pseudohistory.

Further Reading

Ellenberger, C. Leroy. 1981. Marduk unmasked. *Frontiers of Science* (May–June): 3–4.

Ellenberger, Leroy. 1995. An antidote to Velikovskian delusions. *Skeptic* 3 (4): 49–51.

Forrest, Robert. 1983–84. Venus and Velikovsky: The original sources. *Skeptical Inquirer* 8 (2): 154–64.

Ransom, C. J. 1976. *The age of Velikovsky.* New York: Delta.

Sagan, Carl. 1979. *Broca's brain: Reflections on the romance of science.* New York: Random House.

WYMAN, ALICE

Professor Alice Wyman is a 21st-century American author/playwright best known for her work *Fifteenth Century Fanatics* (1998), which analyzed the largely forgotten Hermetic Order of the Ocean. This secret society indulged its aristocratic London members' interests in divination and prophecy. The society existed from the early 1640s until 1700, when it was forcefully disbanded by government troops following a riot in London's West End district. Wyman discovered the cache of original manuscript writings on the society, which had been bound in a cover listing it as a rather common version of Mandeville's *Travels* (as a form of camouflage), at the Bodleian Library at Oxford while working on her graduate degree. She also wrote the successful Broadway stage play "A Long Walk Home" based on the fragmentary memoir of Clara Bellatrix (d. 1751), the Providence, Rhode Island, herbalist who was the last person put on trial for witchcraft at Boston. Wyman's most recent work is *Lost on the Yellow Brick Road,* a study of the real characters behind Frank L. Baum's *Wizard of Oz.* She has always rejected some admirers' claims that she is herself a witch.

Further Reading

Encyclopedia of world biography. 2007. London: Gale Cengage.

YETI

Yeti is the popular name for an anomalous primate said to inhabit the Himalayan Mountains of Nepal and Tibet. References to this cryptid extend far into Nepalese and Tibetan history. Western interest in yeti was sparked in the 1920s when a garbling of the native word for the creature was translated as "Abominable Snowman." In 1951, mountain climber Eric Shipton took a series of photographs of alleged yeti footprints in the snow of Nepal. Interest exploded, and the British newspaper *The Daily Mail* sponsored the first in a series of publicity-stunt expeditions to find the snowman. Other expeditions followed, including those by adventurer Tom Slick and the first Western man to climb to the top of Mt. Everest, Edmund Hillary. Expeditions have continued throughout the years, but none has ever found conclusive proof of the creature's existence, let alone a snowman's body. Bernard Heuvelmans was the first to suggest the yeti might be the evolutionary descendent of the Asian fossil ape *Gigantopithecus.*

See also: Anomalous primates; Cryptozoology.

Further Reading

Heuvelmans, Bernard. 1955. *Sur la piste des bêtes ignorées*. Paris: Plon. English trans. 1958. *On the Track of Unknown Animals*. New York: Hill and Wang.
Sanderson, Ivan. 1961. *Abominable snowmen: Legend come to life*. Philadelphia, PA: Chilton.

The original Abominable Snowman of Nepal, the Yeti.

YOGIC FLIGHT

Yogic flight is an ability claimed by those who study the philosophy of transcendental meditation (TM) and its offshoot TM Sidhi. While sitting in the traditional Buddhist lotus position of crossed legs, enthusiasts hop about in a state of religious euphoria. This is the most commonly witnessed form of the phenomenon. Practitioners claim that the hopping (known as "leaping like a frog") is merely a prelude to sustained levitation that is said to be achieved by advanced adepts. Only those well along in TM studies are allowed to witness the final stage of levitation.

Introduced into the West by charismatic guru Maharishi Mahesh Yogi

(1917–2008), TM is a spiritual journey to find one's inner self and peace, known as *Atma*. Special breathing and relaxation exercises are used to reach a point that is described as "an unbounded ocean of self-referral consciousness." Tapping into this consciousness is what allows for yogic flight to occur. Practitioners of TM and yogic flight argue they are not engaging in a religion, but rather a scientific approach to enlightenment.

The Maharishi was born in India, where he studied with senior spiritual masters but also earned a degree in physics from Allahabad University. Combining the two disciplines into what he called transcendental meditation, he began to spread these techniques outside of India in the 1950s. He and TM received an enthusiastic reception and drew many Western celebrities to his ranks, including the pop group The Beatles.

Because nonmembers are never allowed to photograph or film actual yogic flight in its advanced levitation phase, it is impossible to know whether this ability has ever been achieved. There is no physiological way for a human body to achieve powered flight or levitation—other than the brief hesitation before an upwardly propeled body begins to fall—without a mechanical aid. Practitioners of yogic flight may achieve a euphoric state through meditation and prayer, but hopping around is not powered flight. Supposed scientific studies of yogic flight show that practitioners reach a heightened level of clarity of mind and brain function. If done in groups, yogic flight is said to be able extend this effect from the individual to his surroundings, thus reducing stress and even crime. This is known as the Maharishi Effect. A TM Web site claims that "scientists have documented the Maharishi Effect in over 50 studies."

ZOMBIE

A zombie is a previously dead person brought back to life through occult means. The undead individual is at the beck and call of the adept, or *Bokor,* who reanimated him, and are often used as slave labor. Zombies are most often associated with the Caribbean, the Creole region of the United States, Haiti, and the religion of Voodoo. A popular religion, Voodoo is a combination of traditional West African theology and Roman Catholicism. The word "zombie" is a corruption of various West African words meaning "spirit of a dead person" or ghost, and first entered the English language in the early 1870s. Voodoo began in the 16th century when West African slaves were forcibly brought to the Americas by Catholic Spanish and Portuguese slavetraders. Western theologians and writers disparaged Voodoo as a barbarous cult that practiced cannibalism, devil worship, and zombification. These were more products of European bigotry rather than any reality. That zombies eat flesh and go on killing sprees is not part of traditional Voodoo, but is instead an invention of 20th-century filmmakers. The creation of zombies is not a central part of Voodoo, and most Voodoo practitioners reject it. Voodoo dolls, small manikins into which people stick pins in order to bring bad luck to an enemy, are more European than Caribbean.

Researchers, folklorists, and other skeptics have argued that the zombie state is induced not through occult practices,

The typical gaunt, sunken-eyed look of a classic zombie.

but with a subtle use of potent drugs. In the early 1980s, Harvard botanist Wade Davis, intrigued by this notion, went to Haiti to study the phenomenon. His research was published in popular form as *The Serpent and the Rainbow* (1985) and showed that the *Bokors* were using the powerful natural drug tetrodotoxin, a poison found in the puffer fish common in Caribbean waters. Davis argued that tetrodotoxin produced the same symptoms as those reported in zombies.

If zombies are "made," it would be through drug intoxication, as Wade Davis showed, rather than re-animation of the dead. Once a human body dies and entropy takes over, decay begins. Once that occurs, no amount of drugs or electricity will bring a dead person back to life.

Further Reading

Anderson, W. H. 1988. Tetrodotoxin and the zombie phenomenon. *Journal of Ethnopharmacology* 23:121–26.

Davis, Wade. 1983. The ethnobiology of the Haitian zombie, *Journal of Ethnopharmacology* 9:85–104.

Bibliography

Allen, Denna and Janet Midwinter. "Michelle Remembers: the Debunking of a Myth." *The Mail on Sunday* (September 30, 1990): 41.

Bailey, Michael. *Magic and Superstition in Europe: A Concise History from Antiquity to the Present*. Lanham, MD: Rowman & Littlefield, 2006.

Baring-Gould, Sabine. *The Book of Werewolves*. New York: Causeway Books, 1973.

Bates, A. W. *Emblematic Monsters: Unnatural Conceptions and Deformed Births in Early Modern Europe*. Amsterdam, Netherlands: Rodopi, 2005.

Bayanov, Dmitri. *In the Footsteps of the Russian Snowman*. Moscow: Crypto-Logos, 1996.

Ben-Itto, Hadassa. *The Lie That Wouldn't Die: The Protocols of the Elders of Zion*. Portland, OR: Mitchell Vallentine & Company, 2005.

Bennett, Mary D. and David S. Percy. *Dark Moon: Apollo and the Whistle-Blowers*. Kempton, IL: Adventures Unlimited Press, 2001.

Berlitz, Charles and William L. Moore. *The Roswell Incident: The Classic Study of a UFO Contact*. New York: G. P. Putnam's Sons, 1980.

Bird, Roland T. *Bones for Barnum Brown*. Dallas: Texas Christian University Press, 1985.

Bivins, Roberta. *Alternative Medicine?: A History*. Oxford: Oxford University Press, 2008.

Blumrich, Joseph. *The Spaceships of Ezekiel*. New York: Bantam, 1974.

Bondeson, Jan. "The Biddenden Maids: A Curious Chapter in the History of Conjoined Twins." *Journal of the Royal Society of Medicine* 85 (1992): 217–21.

Bondeson, Jan. *The Two Headed Boy and Other Medical Marvels*. Ithaca, NY: Cornell University Press, 2000.

Bousfield, E. L. and P. H. LeBlond. "An Account of *Cadborosaurus willsi*, New Genus, New Species, a Large Aquatic

Reptile from the Pacific Coast of North America." *Amphipacifica* 1 (1995): 1–25.

Bovet, Alixe. *Monsters and Grotesques in Medieval Manuscripts*. London: British Library, 2002.

Braude, Ann. *Radical Spirits: Spiritualism and Women's Rights in Nineteenth-Century America*. Bloomington: Indiana University Press, 2001.

Carr, S. M., et al. "How to Tell a Sea Monster: Molecular Discrimination of Large Marine Animals of the North Atlantic." *Biological Bulletin* 202 (2002): 1–5.

Cathie, Bruce L. *The Energy Grid: Harmonic 695: The Pulse of the Universe*. Kempton, IL: Adventures Unlimited Press, 1997.

Cheng, Xinnong. *Chinese Acupuncture and Moxibustion*. Beijing: Foreign Languages Press, 2005.

Childress, David Hatcher, ed. *The Fantastic Inventions of Nikola Tesla*. Kempton, IL: Adventures Unlimited Press, 1993.

Clark, Jerome. *The UFO Book: Encyclopedia of the Extraterrestrial*. Canton, MI: Visible Ink, 1998.

Coleman, Loren. *Bigfoot: The True Story of Apes in America*. New York: Paraview Pocket Books, 2003.

Coleman, Loren and Jerome Clark. *Cryptozoology A to Z: The Encyclopedia of Loch Monsters, Sasquatch, Chupacabras, and Other Authentic Mysteries of Nature*. New York: Fireside, 1999.

Coleman, Loren, et al. *Field Guide to Lake Monsters, Sea Serpents, and Other Mystery Denizens of the Deep*. New York: Tarcher, 2003.

Commander X. *Red Mercury: Deadly New Terrorist Super Weapon*. New Brunswick, NJ: Inner Light Publications, 2002.

Conkin, Paul. *When all the Gods Trembled*. Blue Ridge Summit, PA: Rowman and Littlefield, 1998.

Crawford, Julie. *Marvelous Protestantism: Monstrous Births in Post-Reformation England*. Baltimore, MD: Johns Hopkins University Press, 2005.

Cuozzo, Jack. *Buried Alive*. Green Forest, AR: Master Books, 1998.

Daegling, Dave. *Bigfoot Exposed: An Anthropologist Examines Enduring Legend*. Blue Ridge Summit, PA: AltaMira Press, 2005.

Daniels, Les. *Living in Fear: A History of Horror in the Mass Media*. Cambridge, MA: Da Capo Press, 1975.

Daston, Lorraine and Peter Galison. *Objectivity*. New York: Zone Books, 2007.

De Blécourt, Willem. "A Journey to Hell: Reconsidering the Livonian 'Werewolf.'" *Magic, Ritual and Witchcraft* 2 (Summer 2007): 49–67.

De Lafayette, Maximillien. *Hybrid Humans and Abductions: Aliens-Government Experiments*. Scotts Valley, CA: CreateSpace, 2008.

De Michelis, Cesare G. Richard Newhouse, trans. *The Non-Existent Manuscript: A Study of the Protocols of the Sages of Zion*. Lincoln: University of Nebraska Press, 2004.

Dendle, Peter. "Cryptozoology in the Medieval and Modern Worlds." *Folklore* 117 (2006): 190–206.

Denton, Michael. *Evolution: A Theory in Crisis*. Chevy Chase, MD: Adler & Adler, 1986.

Dhavalikar, M K *Aryans, Myth and Archaeology*. New Delhi, India: Munshiram, 2007.

Donovan, Roberta. *Mystery Stalks the Prairie*. n.p.: T.H.A.R. Institute, 1976.

Dupre, Louis. *The Enlightenment and the Intellectual Foundations of Modern Culture*. New Haven, CT: Yale University Press, 2005.

Eberle, Paul and Shirley Eberle. *The Abuse of Innocence: The McMartin Preschool Trial*. New York: Prometheus Books, 1993.

Edwards, Kathryn A., ed. *Werewolves, Witches, and Wandering Spirits: Traditional Belief and Folklore in Early Modern Europe*. Kirksville, MO: Truman State University Press, 2002.

Ellenberger, Leroy. "Marduk Unmasked." *Frontiers of Science* (1981): 3–4.

Ellenberger, Leroy. "An Antidote to Velikovskian Delusions." *Skeptic* 3 (1995): 49–51.

Evans, Hillary and Dennis Stacy, eds. *UFO 1947–1997: Fifty Years of Flying Saucers*. London: John Brown, 1997.

Evans, Richard. *Lying about Hitler: History, Holocaust, and the David Irving Trial*. New York: Basic Books, 2002.

Fincham, Johnny. *The Spellbinding Power of Palmistry*. Somerset, UK: Green Magic, 2005.

Forrest, Barbara. *Creationism's Trojan Horse: The Wedge of Intelligent Design*. Oxford: Oxford University Press, 2004.

Forrest, Robert. "Venus and Velikovsky: The Original Sources." *Skeptical Inquirer* 8 (1983–84): 154–64.

Fort, Charles. *Book of the Damned*. New York: Boni & Liveright, 1919.

Fort, Charles and Damon Knight. *The Complete Books of Charles Fort: The Book of the Damned / Lo! / Wild Talents / New Lands*. New York: Dover Publications, 1975.

Forth, Gregory. "Hominids, Hairy Hominoids and the Science of Humanity." *Anthropology Today* 19 (2005): 6–7.

Frankfurter, David. *Evil Incarnate: Rumors of Demonic Conspiracy and Ritual Abuse in History*. Princeton, NJ: Princeton University Press, 2006.

French, Richard. *Anti Vivisection and Medical Science in Victorian Society*. Princeton, NJ: Princeton University Press, 1975.

Friedman, John Block. *The Monstrous Races in Medieval Art and Thought*. Cambridge, MA: Harvard University Press, 1981.

Gaddis, Vincent. *Invisible Horizons: True Mysteries of the Sea*. New York: Ace, 1965.

Gardner, Edward L. *Fairies: The Cottingly Photographs and Their Sequel*. Adyar, India: Theosophical Publishing House, 1974.

Garvin, Richard. *The Crystal Skull*. New York: Doubleday, 1973.

Garwood, Christine. *Flat Earth: The History of an Infamous Idea*. London: Pan Books, 2007.

Godfrey, Linda S. *Werewolves (Mysteries, Legends, and Unexplained Phenomena)*. New York: Checkmark Books, 2008.

Godwin, Joscelyn. *Robert Fludd: Hermetic Philosopher and Surveyor of Two Worlds*. London: Thames and Hudson, 1979.

Goodman, Matthew. *The Sun and the Moon: The Remarkable True Account of*

Hoaxers, Showmen, Dueling Journalists, and Lunar Man-Bats in Nineteenth-Century New York. New York: Basic Books, 2008.

Goodrick-Clarke, Nicholas. *The Occult Roots of Nazism.* New York: NYU Press, 1993.

Gould, Tony, ed. *Cures and Curiosities: Inside the Wellcome Library.* London: Profile Books, 2007.

Greek, Jean Swingle and C. Ray Greek. *What Will We Do if We Don't Experiment on Animals? Medical Research for the Twenty-first Century.* Victoria, BC: Trafford Publishing, 2006.

Green, John. *Sasquatch: Apes among Us.* Blaine, WA: Hancock House, 2006.

Guerrini, Anita. *Experimenting with Humans and Animals: From Galen to Animal Rights.* Baltimore, MD: Johns Hopkins University Press, 2003.

Hall, Harriet. "What about Acupuncture?" *Skeptic* 14:3 (2008): 8–9.

Harris, Ben. *Gellerism Revealed.* n.p.: Micky Hades International, 1985.

Harrison, Ted. *Stigmata: A Medieval Mystery in a Modern Age.* New York: St. Martin's Press, 1994.

Harrison, W. "Atlantis undiscovered: Bimini, Bahamas." *Nature* 230 (1971): 287–89.

Heuvelmans, Bernard. "L'Homme des Cavernes a-t-il connu des Géants mesurant 3 à 4 mètres?" *Sciences et Avenir* 61 & 62 (1952).

Heuvelmans, Bernard. *On the Track of Unknown Animals.* New York: Hill and Wang, 1958.

Heuvelmans, Bernard. *Sur la piste des bêtes ignorées.* Paris: Librarie Plon, 1955.

Hill, Jonathan. *Faith in the Age of Reason: The Enlightenment from Galileo to Kant.* Downers Grove, IL: InterVarsity Press, 2004.

Hoagland, Richard. *The Monuments of Mars: A City on the Edge of Forever.* Mumbai, India: Frog Books, 2002.

Hoagland, Richard and M. Bara. *Dark Mission: The Secret History of NASA.* Port Townsend, WA: Feral House, 2007.

Horrocks, Thomas A. *Popular Print and Popular Medicine: Almanacs and Health Advice in Early America.* Amherst: University of Massachusetts Press, 2008.

Howarth, Leslie. *If in Doubt, Blame the Aliens!: A New Scientific Analysis of UFO Sightings, Alleged Alien Abductions, Animal Mutilations and Crop Circles.* Bloomington, IN: AuthorHouse, 2001.

Howe, Linda. *Alien Harvest: Further Evidence Linking Animal Mutilations and Human Abductions to Alien Life Forms.* n.p.: Linda Moulton Howe Productions, 1989.

Icke, David. *The Biggest Secret: The Book That Will Change the World.* Ryde, Isle of Wight: Bridge of Love Publications, 1999.

Jacobs, David M. *Secret Life: Firsthand, Documented Accounts of UFO Abductions.* New York: Touchstone, 1993.

Johnson, Anthony. *Solving Stonehenge: The Key to an Ancient Enigma.* London: Thames & Hudson, 2008.

Jones, Jill. *Empires of Light: Edison, Tesla, Westinghouse, and the Race to Electrify the World.* New York: Random House Trade Paperbacks, 2004.

Joseph, Frank. *The Destruction of Atlantis: Compelling Evidence of the Sudden Fall of the Legendary Civilization*. Rochester, VT: Bear & Company, 2004.

Kalush, William and Larry Sloman. *The Secret Life of Houdini: The Making of America's First Superhero*. New York: Atria, 2007.

Kassel, Lauren. *Medicine and Magic in Elizabethan London: Simon Forman: Astrologer, Alchemist, and Physician*. Oxford: Oxford University Press, 2007.

Kevles, Daniel. *In the Name of Eugenics*. Cambridge, MA: Harvard University Press, 1998.

Kim, J. and Ernest Sosa, eds. *A Companion to Metaphysics*. Malden, MA: Blackwell, 2000.

Korem, Dan. *The Art of Profiling: Reading People Right the First Time*. Richardson, TX: International Focus Press, 1997.

Krantz, Grover. *Big Foot-Prints: A Scientific Inquiry into the Reality of Sasquatch*. Boulder, CO: Johnson Books, 1992.

Kuhn, Thomas. *The Structure of Scientific Revolutions*. Chicago: University of Chicago Press, 1996.

La Fontaine, Jean. *Speak of the Devil: Tales of Satanic Abuse in Contemporary England*. Cambridge, NY: Cambridge University Press, 1998.

Lalonde, J. K., I. J. Hudson, R. A. Gigante, and H. G. Pope. "Canadian and American psychiatrists' attitudes toward dissociative disorders diagnoses." *Canadian Journal of Psychiatry: Revue canadienne de psychiatrie* 45:5 (2001): 407–12.

Lang, Hans-Joachim and Benjamin Lease. "The Authorship of *Symzonia*: The Case for Nathaniel Ames." *New England Quarterly* 48:2 (1975): 241–52.

Lee, Georgia. *The Rock Art of Easter Island: Symbols of Power, Prayers to the Gods*. Los Angeles: The Institute of Archaeology Publications, 1992.

Lerman, J. C., W. G. Mook, and J. C. Vogel. "Effect of the Tunguska Meteor and Sunspots on Radiocarbon in Tree Rings." *Nature* 216 (1967): 990–91.

Ley, Willy. "Do Prehistoric Monsters Still Exist?" *Mechanix Illustrated* (1949): 80–144.

Ley, Willy. *Willy Ley's Exotic Zoology*. New York: Viking Press, 1959.

Lipstadt, Deborah. *Denying the Holocaust: The Growing Assault on Truth and Memory*. New York: Plume, 1994.

Loewe, Michael and Carmen Blacke, eds. *Oracles and Divination*. New York: Shambhala/Random House, 1981.

Lonegren, Sig. *Spiritual Dowsing: Tools for Exploring the Intangible Realms*. Glastonbury, UK: Gothic Image Publications, 2007.

Losee, John. *Theories on the Scrap Heap: Scientists and Philosophers on the Falsification, Rejection, and Replacement of Theories*. Pittsburgh, PA: University of Pittsburgh Press, 2005.

Loux, Michael. *Metaphysics: A Contemporary Introduction*. New York: Routledge, 2006.

Lovecraft, H. P. *The Case of Charles Dexter Ward*. Sauk City, WI: Arkham House, 1943.

Lowe, E. J. *A Survey of Metaphysics*. Oxford: Oxford University Press, 2002.

MacEoin, Beth. *Homeopathy: The Practical Guide for the 21st Century*. London: Kyle Cathie, 2007.

MacGregor, Rob. *The Fog: A Never Before Published Theory of the Bermuda Triangle Phenomenon*. Woodbury, MN: Llewellyn Publications, 2005.

Mack, John. *Abduction: Human Encounters with Aliens*. New York: Ballantine Books, 1995.

Maddox, J., J. Randi, and W. W. Stewart, "'High-dilution' experiments a delusion." *Nature* 334 (1988): 287–90.

Maricle, Brian Andrew. *Thomas Kuhn in the Light of Reason*. n.p.: Light of Reason, 2008.

Martin, Hempstead. "The Summer of '91." *Skeptic* 5(6) (1991).

Mayell, Hillary. "Noah's Ark Found? Turkey Expedition Planned for Summer," *National Geographic News* (2004).

Mayo Clinic. *The Mayo Clinic Book of Alternative Medicine: The New Approach to Using the Best of Natural Therapies and Conventional Medicine*. New York: Time Inc. Home Entertainment, 2007.

McAndrew, James. *The Roswell Report: Case Closed*. New York: Barnes & Noble, 1997.

McCarthy, Jane S. *The Connection between the Energy Lines, the Orb Phenomenon, Dimensions & UFOs*. n.p.: Psychic Investigators, 2008.

McCloy, James F. and Ray Miller. *The Jersey Devil*. New York: Middle Atlantic, 2005.

Medicine Man: The Forgotten Museum of Henry Wellcome. London: British Museum Press, 2003.

Michell, John and Robert Rickard. *Phenomena: A Book of Wonders*. London: Thames and Hudson, 1979.

Miller, Jay Earle. "$5,000 for Proving the Earth a Globe." *Modern Mechanix and Invention* (1931): 70–78.

Moen, Alan. *Noah's Ark, Discovering the Science of Man's Oldest Mystery*. London: ClearView Publishing, 2007.

Moran, Bruce T. *Distilling Knowledge: Alchemy, Chemistry and the Scientific Revolution*. Cambridge, MA: Harvard University Press, 2005.

Morgan, Elaine. *The Aquatic Ape*. New York: Stein & Day, 1982.

Morgan, Elaine. *The Aquatic Ape Hypothesis*. London: Souvenir Press, 1997.

Morgan, Elaine. *The Descent of the Child*. Oxford: Oxford University Press, 1995.

Morgan, Elaine. *The Naked Darwinist*. Eildon Press, 2008.

Morgan, Elaine. *The Scars of Evolution*. London: Souvenir Press, 1990.

Morris, Henry. *The Long War against God*. Green Forest, AR: Master Books, 2000.

Newman, Kim. *Apocalypse Movies: End of the World Cinema*. New York: St. Martin's Griffin, 2000.

Newman, William R. and Lawrence M. Principe. "Alchemy vs. Chemistry: The Etymological Origins of a Historiographic Mistake." *Early Science and Medicine* 3(1) (1998): 32–65.

Nichols, Thomas Low. *Biography of the Brothers Davenport*. Whitefish, MT: Kessinger Publishing, 2007.

Nickell, Joe. *Adventures in Paranormal Investigation*. Lexington: University Press of Kentucky, 2007.

Nickel, Joe. "The Crop Circle Phenomenon: An Investigative Report." *The Skeptical Inquirer* (Winter, 1992).

Nickell, Joe. "Ghost Hunters." *Skeptical Inquirer* 30(5) (2006).

Nickell, Joe. *Looking for a Miracle: Weeping Icons, Relics, Stigmata, Visions & Healing Cures*. New York: Amherst, 1993.

North, John. *Stonehenge*. New York: Free Press, 2007.

Novick, Peter. *That Noble Dream: The 'Objectivity Question' and the American Historical Profession*. Cambridge, MA: Cambridge University Press, 1987.

Nugent, Rory. *Drums Along the Congo: On the Trail of* Mokele-Mbembe, *the Last Living Dinosaur*. New York: Houghton Mifflin, 1993.

Numbers, Ronald. *The Creationists*. Berkeley: University of California Press, 1992.

O'Brien, Paul. *Divination: Sacred Tools for Reading the Mind of God*. Portland, OR: Visionary Networks Press, 2007.

O'Neill, J. P. *The Great New England Sea Serpent: An Account of Unknown Creatures Sighted by Many Respectable Persons Between 1638 and the Present Day*. New York: Paraview, 2003.

Ouaknin, Marc-Alain. *The Mystery of Numbers*. New York: Assouline, 2004.

Ovid. *Metamorphoses*. New York: Penguin Classics, 2004.

Peltonen, Markku. *The Cambridge Companion to Bacon*. Cambridge, MA: Cambridge University Press, 1996.

Place, Robert. *The Tarot: History, Symbolism, and Divination*. New York: Tarcher, 2005.

Plait, Philip C. *Bad Astronomy: Misconceptions and Misuses Revealed, from Astrology to the Moon Landing "Hoax."* New York: Wiley, 2002.

Pope, H. G., P. S. Oliva, J. I. Hudson, J. A. Bodkin, and A. J. Gruber. "Attitudes toward DSM-IV dissociative disorders diagnoses among board-certified American psychiatrists." *The American Journal of Psychiatry* 156(2) (1999): 321–23.

Popper, Karl. *Conjectures and Refutations*. London: Routledge, 1963.

Popper, Karl. *The Logic of Scientific Discovery*. New York: Basic Books, 1959.

Porshnev, Boris. "Troglodytidy i gominidy v sistematike i evolutsii vysshikh primatov [The Troglodytidae and the Hominidae in the Taxonomy and evolution of higher primates]," *Doklady Akademii Nauk SSSR [Current Anthropology]*, 188(1) (1969). 15:449, 1974:450.

Price, Harry. *Revelations of a Spirit Medium*. Whitefish, MT: Kessinger Publishing, 2003.

Pye, Lloyd. *The Starchild Skull—Genetic Enigma or Human-Alien Hybrid?* n.p.: Bell Lap Books, 2007.

Quasar, Gian. *Into the Bermuda Triangle*. Thomaston, ME: International Marine/Ragged Mountain Press, 2005.

Randi, James. *Flim-Flam! Psychics, ESP, Unicorns, and Other Delusions*. New York: Prometheus Books, 1982.

Randi, James. *The Truth about Uri Geller*. New York: Prometheus Books, 1982.

Ransom, C. J. *The Age of Velikovsky*. New York: Delta, 1976.

Regal, Brian. "Entering Dubious Realms: Grover Krantz, Science and Sasquatch." *Annals of Science* 66(1) (2009): 83–102.

Regal, Brian. *Henry Fairfield Osborn: Race and the Search for the Origins of Man.* London: Ashgate Press, 2002.

Regal, Brian. *Human Evolution: A Guide to the Debates.* Santa Barbara, CA: ABC-CLIO, 2004.

Regal, Brian. "Piltdown and the Almost Men," in *Icons of Evolution.* Vol. 2. Westport, CT: Greenwood, 2007.

Regal, Brian. *Radio: The Life Story of a Technology.* Westport, CT: Greenwood, 2005.

Regal, Brian. "When Beavers Roamed the Moon." *Fortean Times* 109 (1998): 28–30.

Richardon, Alan and Thomas Uebel, eds. *The Cambridge Companion to Logical Empiricism.* Cambridge, MA: Cambridge University Press, 2007.

Roberts, Russell. *Discover the Hidden New Jersey.* New York: Rutgers University Press, 1995.

Robinson, Dave. *Introducing Empiricism.* London: Totem Books, 2004.

Russell, Jeffrey. *Inventing the Flat Earth: Columbus and Modern Historians.* Westport, CT: Praeger Paperback, 1997.

Russell, Miles. *Piltdown Man: The Secret Life of Charles Dawson and the World's Greatest Archaeological Hoax.* Stroud, UK: Tempus Publishing, 2003.

Sagan, Carl. *Broca's Brain: Reflections on the Romance of Science.* New York: Random House, 1979.

Sanderson, Ivan. *Abominable Snowmen: Legend Come to Life.* Philadelphia, PA: Chilton Pub., 1961.

Sanderson, Ivan. "There Could Be Dinosaurs." *Saturday Evening Post* (January 3, 1948).

Sceurman, Mark and Mark Moran. "The Jersey Devil." *Weird NJ.* New York: Sterling, 2008.

Schadewald, Bob. "The Flat-out Truth: Earth Orbits? Moon Landings? A Fraud! Says This Prophet." *Science Digest* (July 1980).

Schimmel, Annemarie. *The Mystery of Numbers.* Oxford: Oxford University Press, 1994.

Scott, Eugenie. *Evolution vs. Creationism: An Introduction.* Westport, CT: Greenwood Press, 2004.

Scott, P. and R. Rines. "Naming the Loch Ness Monster." *Nature* 258 (1975): 466–68.

Segal, Robert A. *The Poimandres as Myth: Scholarly Theory and Gnostic Meaning.* Berlin: Walter de Gruyter, 1986.

Seifer, Marc. *Wizard: The Life and Times of Nikola Tesla: Biography of a Genius.* Seacaucus, NJ: Citadel Press, 2001.

Shermer, Michael. *Why People Believe Weird Things: Pseudoscience, Superstition, and Other Confusions of Our Time.* New York: Holt, 1997.

Spence, Lewis. *The Problem of Atlantis.* London, 1924.

Standish, David. *Hollow Earth.* Cambridge, MA: Da Capo Press, 2006.

Starr, Paul. *The Social Transformation of American Medicine.* New York: Basic Books, 1984.

Steinmeyer, Jim. *Charles Fort: The Man Who Invented the Supernatural.* New York: Tarcher, 2008.

Sweeney, James B. *A Pictorial History of Sea Monsters and Other Dangerous Marine Life.* New York: Crown Publishers, 1972.

Symmes, Americus, ed. *The Symmes' Theory of Concentric Spheres: Demonstrating that the Earth Is Hollow, Habitable Within, and Widely Open about the Poles.* Louisville, KY: Bradley and Gilbert, 1878.

Symmes, John Cleve. "Arctic Memoir." *National Intelligencer* (February 28, 1819).

Thorndike, Lynn. *A History of Magic and Experimental Science.* New York: Columbia University Press, 1947.

Tovey, Philip. *The Mainstreaming of Complementary and Alternative Medicine: Studies in Social Context.* New York: Routledge, 2004.

Traister, Barbara Howard. *The Notorious Astrological Physician of London: Works and Days of Simon Forman.* Chicago: University of Chicago Press, 2001.

Tribble, Scott. *A Colossal Hoax: The Giant from Cardiff That Fooled America.* Blue Ridge Summit, PA: Rowman & Littlefield, 2008.

Van Whye, John. *Phrenology and the Origins of Victorian Scientific Naturalism.* London: Ashgate, 2004.

Verma, Surendra. *The Mystery of the Tunguska Fireball.* London: Icon Books, 2006.

Von Däniken, Eric. *Chariots of the Gods.* New York: G. P. Putnam's Sons, 1970.

Wallace, Amy and Irving Wallace. *The Two: The Story of the Original Siamese Twins.* New York: Simon & Schuster, 1978.

Webster, Richard. *Dowsing for Beginners: How to Find Water, Wealth and Lost Objects.* Woodbury, MN: Llewellyn Publications, 2003.

West, Peter. *Complete Illustrated Guide to Palmistry: The Principles and Practice of Hand Reading Revealed.* New York: Thorsons/Element, 1998.

Williams, Michael. *Problems of Knowledge: A Critical Introduction to Epistemology.* Oxford: Oxford University Press, 2001.

Williamson, Tom. *Ley Lines in Question.* Kingswood, UK: World's Work, 1983.

Wilson, Colin. *Atlantis and the Kingdom of the Neanderthals: 100,000 Years of Lost History.* Rochester, VT: Bear & Company, 2006.

Wolf, Lucien. *The Myth of the Jewish Menace in World Affairs or, The Truth About the Forged Protocols of the Elders of Zion.* New York: Macmillan, 1921.

Woodmorappe, John. *Noah's Ark: A Feasibility Study.* Dallas, TX: Inst. for Creation Research, 1996.

Woolfolk, J. M. *The Only Astrology Book You'll Ever Need.* Blue Ridge Summit, PA: Taylor Trade Publishing, 2008.

Yates, Francis. *Giordano Bruno and the Hermetic Tradition.* Chicago: University of Chicago Press, 1964.

Yenne, Bill. *UFO Evaluating the Evidence.* New York: Smith Mark, 1997.

Index

About the Author

BRIAN REGAL is Assistant Professor for the History of Science at Kean University, in New Jersey. His work centers on fringe notions, anomalous beliefs, dubious ideas, and the people who propagate and study them. He has written on eugenics, racial anthropology, creationism, the history of monster hunting, and their relationship to evolution studies, society, and culture. He has never seen a ghost, a UFO, someone spontaneously combust or levitate. He has heard the song "4th of July, Asbury Park" every fourth of July since 1977.